Studies in Applied Mechanics 50

Rotating Shell Dynamics

Studies in Applied Mechanics

20. Micromechanics of Granular Materials (Satake and Jenkins, Editors)
21. Plasticity. Theory and Engineering Applications (Kaliszky)
22. Stability in the Dynamics of Metal Cutting (Chiriacescu)
23. Stress Analysis by Boundary Element Methods (Balas, Sládek and Sládek)
24. Advances in the Theory of Plates and Shells (Voyiadjis and Karamanlidis, Editors)
25. Convex Models of Uncertainty in Applied Mechanics (Ben-Haim and Elishakoff)
28. Foundations of Mechanics (Zorski, Editor)
29. Mechanics of Composite Materials - A Unified Micromechanical Approach (Aboudi)
31. Advances in Micromechanics of Granular Materials (Shen, Satake, Mehrabadi, Chang and Campbell, Editors)
32. New Advances in Computational Structural Mechanics (Ladevèze and Zienkiewicz, Editors)
33. Numerical Methods for Problems in Infinite Domains (Givoli)
34. Damage in Composite Materials (Voyiadjis, Editor)
35. Mechanics of Materials and Structures (Voyiadjis, Bank and Jacobs, Editors)
36. Advanced Theories of Hypoid Gears (Wang and Ghosh)
37A. Constitutive Equations for Engineering Materials
     Volume 1: Elasticity and Modeling (Chen and Saleeb)
37B. Constitutive Equations for Engineering Materials
     Volume 2: Plasticity and Modeling (Chen)
38. Problems of Technological Plasticity (Druyanov and Nepershin)
39. Probabilistic and Convex Modelling of Acoustically Excited Structures (Elishakoff, Lin and Zhu)
40. Stability of Structures by Finite Element Methods (Waszczyszyn, Cichoń and Radwańska)
41. Inelasticity and Micromechanics of Metal Matrix Composites (Voyiadjis and Ju, Editors)
42. Mechanics of Geomaterial Interfaces (Selvadurai and Boulon, Editors)
43. Materials Processing Defects (Ghosh and Predeleanu, Editors)
44. Damage and Interfacial Debonding in Composites (Voyiadjis and Allen, Editors)
45. Advanced Methods in Materials Processing Defects (Predeleanu and Gilormini, Editors)
46. Damage Mechanics in Engineering Materials (Voyiadjis, Ju and Chaboche, Editors)
47. Advances in Adaptive Computational Methods in Mechanics (Ladevèze and Oden, Editors)
48. Stability of Nonlinear Shells - On the Example of Spherical Shells (Shilkrut, Editor)
49. Introduction to Hydrocodes (Zukas)

General Advisory Editor to this Series:
Professor Isaac Elishakoff, Center for Applied Stochastics Research, Department of Mechanical Engineering, Florida Atlantic University, Boca Raton, FL, U.S.A.
elishako@fau.edu

Studies in Applied Mechanics 50

Rotating Shell Dynamics

Hua Li
Computational MEMS Division,
Institute of High Performance Computing,
Singapore

K.Y. Lam
Department of Mechanical Engineering,
National University of Singapore,
Singapore

T.Y. Ng
School of Mechanical and Production Engineering,
Nanyang Technological University,
Singapore

ELSEVIER

2005

Amsterdam – Boston – Heidelberg – London – New York – Oxford
Paris – San Diego – San Francisco – Singapore – Sydney – Tokyo

ELSEVIER B.V.
Sara Burgerhartstraat 25
P.O. Box 211, 1000 AE
Amsterdam, The Netherlands

ELSEVIER Inc.
525 B Street
Suite 1900, San Diego
CA 92101-4495, USA

ELSEVIER Ltd.
The Boulevard
Langford Lane, Kidlington,
Oxford OX5 1GB, UK

ELSEVIER Ltd.
84 Theobalds Road
London WC1X 8RR
UK

First edition 2005

Library of Congress Cataloging in Publication Data
Li, Hua.
   Rotating shell dynamics/Hua Li, K.Y. Lam, T.Y. Ng. – 1st ed.
      p. cm. – (Studies in applied mechanics ; 50)
   Includes bibliographical references and index.
   ISBN 0-08-044477-6 (alk. paper)
      1. Shells (Engineering) I. Lam, Khin Yong. II. Ng, T. Y. (Teng Yong) III. Title. IV.
   Series.

   TA660.S5L453 2004
   624.17762–dc22

                                                                    2004057705
A catalog record is available from the Library of Congress.

British Library Cataloguing in Publication Data
A catalogue record is available from the British Library.

ISBN: 0-08-044477-6

⊗ The paper used in this publication meets the requirements of ANSI/NISO Z39.48-1992 (Permanence of Paper).
Printed in The United Kingdom.

Dedicated to Duer, Anne and my parents for their constant encouragement
Hua Li

To Karen, Derrick and Rachel for their support and encouragement
K.Y. Lam

Dedicated first and foremost to GOD, and to my wife, Lay Keow
T.Y. Ng

PREFACE

There are numerous engineering applications for high-speed rotating shell structures which rotate about their symmetric axes, and this is especially so in aerospace industries. Although there are many published books on shell dynamics, almost all of them do not involve the dynamics of *rotating* shells. Physically, the important distinctions between rotating and non-rotating shells of revolution are the presence of Coriolis and centrifugal accelerations as well as the hoop tension arising in rotating shells due to the angular velocities. These effects have significant influences on the dynamic behaviors of the rotating shells. For example, the frequency characteristics of a shell structure are generally determined by the shell geometry, material properties, boundary conditions and externally applied loads. However, when the shell rotates, the structural frequency characteristics are qualitatively altered. This qualitative difference manifests itself in the form of a bifurcation phenomenon in the natural frequency parameters. For a stationary shell of revolution, the vibration of the shell is a standing wave motion. However, when the same shell rotates, the standing wave motion is transformed, and depending on the direction of rotation, backward or forward travelling waves will emerge.

Due to the distinct differences from generic stationary shells, and their wide-ranging engineering applications, a comprehensive study is therefore warranted for the full understanding of rotating shell dynamics. In the latter half of the last decade, such a study was performed by the authors. Much of this research was carried out in the context of graduate research work, with Professor Khin-Yong Lam (current co-author) of the National University of Singapore supervising a group of graduate students in this focus area. The two other current co-authors were then PhD research scholars in this very vibrant group. The idea of collating and systematically documenting this research was first hatched in June 2002 when the authors were attending a scientific meeting in Philadelphia. Amazingly, we acted upon it upon our return to Singapore! The present monograph is written with a view to share the key developments and integral findings of the above research with the research community. In this monograph, a complete theoretical platform detailing of the fundamental theory for rotating shells of revolution is established. Dynamic problems such as free vibration and dynamic stability are examined in detail, for basic shells of revolution such as cylindrical and conical shells. The influences of various parameters on the dynamic behaviors, including rotating speed, boundary condition, initial pressure, geometrical and material properties, wave number, etc., are investigated in various parametric studies.

This work represents the first monographic text fully dedicated to the dynamic behaviors of rotating shell structures. It aims to provide both the casual and interested reader with insights into the special features and intricacies of shell dynamics when rotation is involved, and covers the basic derivation of the dynamic governing equations for rotating shells. Benchmark results for free vibration, critical speed, and parametric resonance are also documented. It is written in as simple a manner as possible, making it informative and easy reading for both the expert and the casual reader. Concurrently it can serve as a rich reference source for scientists and engineers in this area who wish to carry out computational studies such as the modeling and simulation of rotating shell-like components. It will also be invaluable to design engineers in the relevant industries, serving as a useful reference source with benchmark results to compare and verify their experimental data against. Undergraduate students taking advanced aerodynamics or solid mechanics courses, which involve rotating shell dynamics, may also find the chapter on the fundamental theoretical development a useful and handy reference.

The authors would like to thank Professor J.N. Reddy for his constant useful and helpful advice over the years, and especially for writing the foreword to this book. Special thanks also go to Mr Chee-Tong Loy and Dr Qian Wu for their invaluable contributions to this research.

Hua LI
Institute of High Performance Computing,
Singapore

Khin-Yong LAM
National University of Singapore,
Singapore

Teng-Yong NG
Nanyang Technological University,
Singapore

FOREWORD

It is my pleasure to write the foreword for this monograph, not only because I have known the present authors and their work for many years, but also because I believe this book contains many important contributions to the dynamics of rotating shell structures, with wide-ranging applications in aerospace as well as mechanical engineering. There are a large number of journal publications and books in the area of shell dynamics, however, only a small proportion of them involve the dynamics of high-speed rotating shells. Furthermore, the dynamic characteristics of the rotating shells are significantly different from those of generic stationary shells.

The mathematics of dynamical theories of structures have more than 300 years of history since Galileo Galilei first studied the variation of the natural frequency of a simple pendulum in the first half of the 17th century. The basic shell theory also has a long history since the Love–Kirchhoff hypotheses were established in the second half of the 19th century. Since then, there have been numerous researchers working on the dynamics of shell structures. To my knowledge, however, this is first monograph specially focusing on the dynamic behavior of shells of revolution rotating about their symmetrical axes. This book provides a complete exposition of the dynamics of shells of revolution, including basic derivations of the governing equations of motion of the rotating shells. Also examined are the influences of Coriolis and centrifugal accelerations, rotating velocity, geometric and physical parameters, boundary conditions, initial stresses, modal wave number, and transverse shear deformation.

This book should serve as a valuable reference to researchers who are interested in the analysis of rotating shells. This book is also highly recommended for design engineers dealing with rotating shell-like structures. Graduate students involved with rotating shell dynamics will also find this a very helpful reference.

J. N. Reddy
Distinguished Professor and
Holder of Oscar S. Wyatt Endowed Chair,
Department of Mechanical Engineering,
Texas A&M University, College Station,
Texas 77843-3123

AUTHORS' BRIEF BIOGRAPHIES

Dr Hua LI received his BSc and MEng degrees in Engineering Mechanics from Wuhan University of Technology, PRC, in 1982 and 1987, respectively. He obtained his PhD degree in Mechanical Engineering from the National University of Singapore in 1999.

From 2000 to 2001, Dr Li was Postdoctoral Associate at the Beckman Institute for Advanced Science and Technology, University of Illinois at Urbana-Champaign. Currently he is Research Scientist *cum* Acting Programme Manager of the MEMS/NEMS Modelling Programme in the Institute of High Performance Computing (IHPC), Singapore.

Dr Li's current research interests include the modelling and simulation of MEMS/NEMS, focussing on responsive hydrogels with BioMEMS applications, development of numerical methodology, dynamics of high-speed rotating shells, and vibration of composites structures. He has over 90 technical publications in these areas, including 60 international refereed journal papers, many of which are well cited. In 2003, he was part of the team which won the Silver Award of HPC Quest — The Blue Challenge — presented by IBM and IHPC.

Prof. Lam Khin YONG obtained his first degree in Mechanical Engineering from Imperial College in 1980 and his Masters and PhD degrees from the Massachusetts Institute of Technology in 1982 and 1985, respectively. He also attended the Advanced Management Program at the Harvard Business School in 2000.

He was instrumental in the establishment of the Institute of High Performance Computing (IHPC) in 1998 and was the Institute's Executive Director from 1998 to 2003. Currently, he is the Executive Director of the Agency for Science, Technology and Research (A*STAR) Graduate Academy.

Prof. Lam's research interest is in Computational Mechanics involving shock and vibration, composites and numerical techniques. He has over 358 technical publications, including 241 international refereed journal papers and has over 960 citations since 1990. For his contributions to Computational Mechanics, he was awarded the National Young Scientist and Engineer Award in 1990 and the Outstanding University Research Award in 1997. In 1998, he led the Underwater Shock Technology Team to win the MINDEF Defence Technology Prize. This was the first time such recognition was given to a team comprising members from the university/research institute.

Dr Teng Yong NG earned his BEng, MEng and PhD degrees in Mechanical Engineering from the National University of Singapore (NUS) in 1992, 1995 and 1998, respectively. He was Research Engineer at the Centre for Computational Mechanics in NUS in 1997 before being appointed as Research Manager in 1998 and establishing the Computational MEMS Division at the Institute of High Performance Computing (IHPC), Singapore. In 2002, Dr Ng left IHPC to take up an appointment as Assistant Professor at the School of Mechanical and Production Engineering in the Nanyang Technological University (NTU), Singapore. He has been with NTU since.

Dr Ng's research interests cover various aspects of Mathematical Modelling and Methodology Development. He has been involved in the development of meshless techniques, smart materials modelling, plate and shell dynamics, and more recently molecular and multiscale modelling. He has over 100 technical publications in these areas, including 80 international refereed journal papers, many of which have been extensively cited. For his contributions to Modelling and Simulation, Dr Ng was nominated finalist for three consecutive years, 2000–2002, by the Singapore National Academy of Science for the prestigious National Young Scientist Award.

ACKNOWLEDGEMENT

The authors would like to sincerely acknowledge the use of figures and tables reproduced from the following sources.

Y. Chen, H.B. Zhao, Z.P. Shen, I. Grieger and B.H. Kröplin. (1993). Vibrations of high speed rotating shells with calculations for cylindrical shells. *Journal of Sound and Vibration*, 160, 137-160.

D. Guo, F.L. Chu and Z.C. Zheng. (2001). The influence of rotation on vibration of a thick cylindrical shell. *Journal of Sound and Vibration*, 242, 487-505.

K.Y. Lam and Hua Li. (1997). Vibration analysis of a rotating truncated circular conical shell. *International Journal of Solids and Structures*, 34, 2183-2197.

K.Y. Lam and Hua Li. (1999a). Influence of boundary conditions on the frequency characteristics of a rotating truncated circular conical shell. *Journal of Sound and Vibration*, 223, 171-195.

K.Y. Lam and Hua Li. (1999b). On free vibration of a rotating truncated circular orthotropic conical shell. *Composites: Part B*, 30, 135-144.

K.Y. Lam and Hua Li. (2000a). Influence of initial pressure on frequency characteristics of a rotating truncated circular conical shell. *International Journal of Mechanical Sciences*, 42, 213-236.

K.Y. Lam and Hua Li. (2000b). Generalized differential quadrature for frequency of rotating multilayered conical shell. *ASCE Journal of Engineering Mechanics*, 126, 1156-1162.

K.Y. Lam, Hua Li, T.Y. Ng and C.F. Chua. (2002). Generalized differential quadrature method for the free vibration of truncated conical panels. *Journal of Sound and Vibration*, 251, 329-348.

K.Y. Lam and C.T. Loy. (1994). On vibrations of thin rotating laminated composite cylindrical shells. *Composites Engineering*, 4, 1153-1167.

K.Y. Lam and C.T. Loy. (1995a). Free vibrations of a rotating multi-layered cylindrical shell. *International Journal of Solids and Structures*, 32, 647-663.

K.Y. Lam and C.T. Loy. (1995b). Analysis of rotating laminated cylindrical shells by different thin shell theories. *Journal of Sound and Vibration*, 186, 23-35.

K.Y. Lam and C.T. Loy. (1998). Influence of boundary conditions for a thin laminated rotating cylindrical shell. *Composite Structures*, 41, 215-228.

K.Y. Lam and Q. Wu. (1999). Vibrations of thick rotating laminated composite cylindrical shells. *Journal of Sound and Vibration*, 225, 483-501.

Hua Li. (2000a). Influence of boundary conditions on the free vibrations of rotating truncated circular multi-layered conical shells. *Composites: Part B*, 31, 265-275.

Hua Li. (2000b). Frequency characteristics of a rotating truncated circular layered conical shell. *Composite Structures*, 50, 59-68.

Hua Li. (2000c). Frequency analysis of rotating truncated circular orthotropic conical shells with different boundary conditions. *Composites Science and Technology*, 60, 2945-2955.

Hua Li and K.Y. Lam. (1998). Frequency characteristics of a thin rotating cylindrical shell using the generalized differential quadrature method. *International Journal of Mechanical Sciences*, 40, 443-459.

Hua Li and K.Y. Lam. (2000). The generalized differential quadrature method for frequency analysis of a rotating conical shell with initial pressure. *International Journal for Numerical Methods in Engineering*, 48, 1703-1722.

Hua Li and K.Y. Lam. (2001). Orthotropic influence on frequency characteristics of a rotating composite laminated conical shell by the generalized differential quadrature method. *International Journal of Solids and Structures*, 38, 3995-4015.

T.Y. Ng. (2003). Erratum to Parametric resonance of a rotating cylindrical shell subjected to periodic axial loads [*Journal of Sound and Vibration* 214 (1998) 513-529]. *Journal of Sound and Vibration*, 263, 705-708.

T.Y. Ng and K.Y. Lam. (1999). Vibration and critical speed of a rotating cylindrical shell subjected to axial loading. *Applied Acoustics*, 56, 273-282.

T.Y. Ng, K.Y. Lam and J.N. Reddy. (1998). Parametric resonance of a rotating cylindrical shell subjected to periodic axial loads. *Journal of Sound and Vibration*, 214, 513-529.

T.Y. Ng, Hua Li and K.Y. Lam. (2003). Generalized differential quadrature for free vibration of rotating composite laminated conical shell with various boundary conditions. *International Journal of Mechanical Sciences*, 45, 567-587.

CONTENTS

Preface .... *vii*

Foreword .... *ix*

Authors' Brief Biographies .... *xi*

Acknowledgement .... *xv*

Chapter 1
Introduction

1.1   Rotating shells of revolution 1

1.2   Historical development of the dynamics of shells 2

1.3   About this monograph 4

Chapter 2
Fundamental Theory of Rotating Shells of Revolution

2.1   Basic considerations and assumptions 7

2.2   Shell kinematic strain–displacement relations 8

2.3   Resultant stress–strain relations in constitutive shell models 11

2.4   Governing equations of motion 16

2.5   Eigenvalue analysis of boundary value problems 24

Chapter 3
Free Vibration of Thin Rotating Cylindrical Shells

3.1   Introduction 27

3.2   Theoretical development: rotating thin cylindrical shell 28

3.3   Numerical implementation 33

    a)   Galerkin's method (characteristic beam functions) 33

    b)   Convergence characteristics and numerical validation 39

3.4   Frequency characteristics 45

    a)   Influence of Coriolis and centrifugal effects 45

    b)   Different thin-shell theories 47

    c)   Influence of rotating velocity 52

    d)   Influence of length and thickness 58

e)   Influence of layered configuration of composites 68

f)   Influence of boundary condition. 75

g)   Discussion on modal wave numbers 84

Appendix A 89

Chapter 4
Free Vibration of Thin Rotating Conical Shells

4.1   Introduction 91

4.2   Theoretical development: rotating conical shell. 92

4.3   Numerical implementation. 113

   a)   Assumed-mode method and generalized differential quadrature. 113

   b)   Convergence characteristics and numerical validation. 137

4.4   Frequency characteristics. 144

   a)   Influence of rotating velocity. 144

   b)   Influence of cone angle. 150

   c)   Influence of length and thickness. 152

   d)   Influence of orthotropy and layered configuration of composites 160

   e)   Influence of boundary condition. 171

   f)   Influence of initial stress 179

   g)   Discussion on wave number 190

Appendix A 195

Appendix B. 198

Chapter 5
Free Vibration of Thick Rotating Cylindrical Shells

5.1   Introduction 201

5.2   Natural frequency analysis by Mindlin shell theory. 202

   a)   Rotating Mindlin shell theory — development 202

   b)   Numerical validation and comparison 207

   c)   Frequency characteristics. 209

5.3   Analysis of vibrational mode by FEM with nonlinear kinematics. 213

   a)   Classification of three-dimensional modes of thick
         rotating cylindrical shells 213

   b)   Numerical implementation 218

   c)   Influence of rotation on frequencies of various three-dimensional
         modes 219

Chapter 6
Critical Speed and Dynamic Stability of Thin Rotating Isotropic Cylindrical Shells

6.1 Introduction 229
6.2 Theoretical development: axially loaded rotating shells. 230
6.3 Numerical implementation. 232
    a) Critical speed analysis. 232
    b) Dynamic stability analysis. 235
6.4 Critical speeds and instability regions. 239
    a) Influence of axial loading on critical speeds. 239
    b) Parametric studies on dynamic stability 242

References 251
Subject Index 261

Chapter 1
Introduction

1.1   Rotating shells of revolution.

A shell is a three-dimensional body that is bounded by two closely spaced curved surfaces, in which the distance between the surfaces is small in comparison with the other dimensions. This distance is very often constant, leading to a shell with constant thickness. A shell has three fundamental identifying features, namely, the reference surface, thickness and edge conditions. Of these, the reference surface is the most significant as it defines the shape of the shell. For shells with material constitution which is isotropic or composite laminated, which are the cases considered in this monograph, the reference surface is usually taken to be the middle surface on which the loci of the points defining this middle surface are equidistant from the two boundary surfaces.

A shell of revolution is defined here as a shell having its middle surface formed by rotating a planar curve 360° about a revolution axis which also lies in this plane. The planar curve is sometimes termed the meridian of the surface. As a result, the shell structure becomes axisymmetrical. A shell can possess just a single axis, such as a rubber hose structure, or it can have multiple axes, such as a fish bowl. Further, it can be open like a roof gutter, or it can be closed like a balloon. Typical shells of revolution include cylindrical, conical, spherical and toroidal shells. Axisymmetric shells of revolution are the most widely encountered shell structures employed in engineering applications. The technical importance is considerable due to the following practical considerations:

1) *Ease of fabrication.* Axisymmetric bodies, such as pipes, bottles and cans, are usually easier to manufacture than bodies with more complex geometries. Standard processes include extrusion, drawing and casting.

2) *Fatigue strength.* Axisymmetric configurations are often optimal in terms of strength-to-weight ratio due to the inherently favorable distribution of the structural material. Recall that the strongest columns and shafts, if wall buckling is ignored, have annular cross-sections.

3)   *Diverse applications.* Hollow axisymmetric bodies can often assume a dual purpose, namely, as the overall structure withstanding external loadings and concurrently acting as the housing shelter for internal components.

Furthermore, it should be noted that there are numerous engineering applications for shells of revolution rotating at high angular velocities about their symmetric longitudinal axes. For example, free-flight sub-munition projectiles rotate at high speeds for an aerodynamically stable flight. Physically, the important differences between the rotating and non-rotating shells of revolution are the Coriolis and centrifugal accelerations as well as the hoop tension arising in rotating shells due to the constant angular velocities. These effects have significant influences on the dynamic behavior of the rotating shells. For example, the frequency characteristics of a shell structure are generally determined by the shell geometry, material properties and boundary conditions. However, when the shell rotates about its symmetric longitudinal axis, the structural frequency characteristics are qualitatively altered and significantly influenced by the rotation.

Mathematically, this qualitative difference manifests itself in the form of a bifurcation phenomenon in the natural frequency parameters. For a stationary shell of revolution, the vibration of the shell is a standing wave motion. However, when the shell rotates, the standing wave motion is transformed and, depending on the direction of rotation, backward or forward waves will emerge. Therefore, the study and understanding of rotating shell dynamics are absolutely necessary and important due to the distinct differences from generic stationary shells and their wide range of engineering applications.

1.2   Historical development of the dynamics of shells.

The widespread engineering applications of rotating shells of revolution in modern industry are becoming ever more evident, and are especially pervasive in the aerospace industry. An example is sub-munitions projectiles or rockets having to rotate at high speeds for an aero-dynamically stable flight. Other examples include advanced gas turbines, high-powered aircraft jet engines, high-speed centrifugal separators, etc. In rotating-shell analysis, flexural vibrations and resulting fatigue fracture under high rotational speeds are of great practical interest. Extensive experimental data have shown that rotors crack even when the rotation-induced stresses are below those supposedly allowed by the strengths of the materials. It is well established now that the factors causing such failures include the detrimental effects of material fatigue and inherent structural defects, and these damaging effects are augmented by the vibrations of the rotating-shell structures. It is thus of practical importance to understand the dynamics of rotating shells of revolution.

The study of the free vibration phenomenon and the development of the associated mathematics possess a very long historical backdrop. Investigation into vibration finds its beginnings with Galileo [1638], who solved, by geometrical means, the dependence of the natural frequency of a simple pendulum on the pendulum length. He then proceeded to make experimental observations on the vibration behavior of strings and plates but, due to the mathematical limitations of his time, he could not offer any analytical treatments. The study of basic shell structures is also very well established. The fundamental theories were first established in the second half of the 19th century. Love [1888] made important contributions to this field. Two assumptions for shells, which he introduced to supplement Kirchhoff's [1876] assumptions from the theory of thin plates, provided the grounds for what came to be known as the Kirchhoff–Love thin-shell theory, which is commonly used even today. This theory is sometimes referred to as Love's first approximation. A more precise theory, which further supplemented the Kirchhoff–Love assumptions by accounting for the influences of transverse shear effects, is due chiefly to Reissner [1944, 1952]. By the time Leissa [1993] completed his monograph on the vibration of shells, the level of mathematics had enabled analytical solutions to be obtained for thin shells with simple geometries but with arbitrary boundary conditions. Also, many alternative thin-shell theories had been proposed by then, such as those of Donnell [1933], Flügge [1960] and Sanders [1959], which were developed through making slight modifications to the Love–Kirchhoff theory. These thin-shell theories, which make use of the Love–Kirchhoff assumptions and which do not take into account the transverse shear effects, are collectively known as *classical* shell theories. Shell theories that account for transverse shear effects, such as those of Reissner [1944, 1952], are collectively known as *refined* shell theories.

The first recorded work on the dynamics of a rotating ring, as a special case of a shell, was by Bryan [1890]. Using the inextensional, shearless deformation and negligible rotatory inertia assumptions, he modeled the free vibration of a spinning ring and it was in this piece of work that the travelling-mode phenomenon was first discovered. Other early works included studies by Brzoska [1953] and Grybos [1961] on the critical speed of isotropic thin-walled drum-type rotors using Love's first approximation, where the critical speeds for a special case of a thin-walled cylinder were obtained. The effects of Coriolis forces were first discussed by DiTaranto & Lessen [1964] for an infinitely long rotating thin-walled isotropic circular cylindrical shell. He concluded that the Coriolis effects have a large influence on the natural frequencies and should always be considered in such analyses. Srinivasan & Lauterbach [1971] later combined both the effects of Coriolis forces and traveling modes in the study of infinitely long, isotropic cylindrical shells. Zohar & Aboudi [1973] later applied both these effects in their investigations on finite-length rotating shells. Other works include Padovan [1973, 1975a,b], who examined the effects of pre-stress on the free vibration of rotating cylinders, as well as the buckling of rotating anisotropic shells.

In these early works, due to the lack of computational power and robust numerical techniques, most of the studies on rotating-shell dynamics were carried out via analytical approaches and were thus restricted to simple cases. Since the advent of high-performance computers, coupled with the development of powerful discrete numerical techniques such as finite elements, shell analysis has taken on a whole new outlook. With these new tools at their disposal, researchers are moving on to solve more and more complicated shell-dynamics problems and these tools currently act as the main drivers for these studies.

1.3   About this monograph.

This monograph provides a comprehensive and systematic study of the dynamics of rotating shells of revolution and is based on the authors' works conducted over the last decade. The monograph is divided into six chapters and each chapter is further divided into sections to better organize the materials. Chapter 1 gives an introduction and provides a historical background on rotating-shell dynamics.

Chapter 2 presents the fundamental theory of the rotating shells of revolution. After stating the basic definitions and hypotheses for the shell problem, the constitutive and geometric equations are given to describe the stress–strain and strain–displacement relations, respectively. The governing equations of motion, based on the classical equilibrium analysis of infinitesimal body and asymptotic expansion of linear approximation, are then derived. To set the groundwork, the generalized eigenvalue problem is summarized for the free-vibration analysis of these rotating shells of revolution.

In Chapter 3, the fundamental theory of the rotating shells of revolution described in Chapter 2 is reduced to that of thin rotating cylindrical shells. A Galerkin-based method employing characteristic beam functions is used to examine the free vibration problem. A wide range of parametric studies, which include both the effects of centrifugal and Coriolis accelerations, are carried out. Effects of boundary conditions on the frequency parameters are also examined. Four different classical thin-shell theories, unified through the use of tracers, are presented and a detailed comparison of the numerical results generated by these different theories is conducted.

Chapter 4 applies the fundamental theory of the rotating shells of revolution developed in Chapter 2 to the free vibration analysis of rotating conical shells. An assumed-mode method based on the generalized differential quadrature is developed here, and numerical validations are carried out to ensure the numerical stability and accuracy of the developed method. A wide range of parametric studies, which include the effects of initial stress on the frequency parameters, are carried out.

In Chapter 5, the rotating Mindlin shell theory is developed for the free vibration analysis of thick rotating cylindrical shells. The theoretical development involves the inclusion of the rotary inertia and Coriolis terms into the first-order shear deformation Mindlin shell theory. In the numerical implementation, an assumed-mode method based on orthogonal polynomials is employed here. After numerical validations are carried out, the frequency characteristics are obtained for both infinitely long and finite-length thick rotating cylindrical shells.

Chapter 6 examines the critical speed of thin rotating cylindrical shells subjected to constant axial loading. Further, the dynamic stability of thin rotating cylindrical shells subjected to harmonic axial loading is also investigated. An assumed mode based on characteristic beam functions is used to formulate the problem. The instability regions are obtained via Bolotin's [1964] method. Detailed parametric studies for the critical speeds and instability regions with respect to rotational speed and axial loading are carried out.

Chapter 2
Fundamental Theory of Rotating Shells of Revolution

2.1   Basic considerations and assumptions.

Throughout this monograph, with the exception of Chapter 5, all analyses and discussions will be with reference to thin shells. As is well known, all classical thin-shell theories find their basis in the Love–Kirchhoff hypotheses, which state that

1.   The thickness of the shell is very small compared with its other dimensions.

2.   The deformation of the shell is small.

3.   The transverse normal stress is negligible.

4.   The normal to the reference surface of the shell remains normal to the deformed surface.

5.   The normal to the reference surface undergoes negligible change in length during deformation.

These five hypotheses, first published in 1892, are sometimes referred to as Love's first approximation for thin shells. The last three hypotheses in fact express Kirchhoff's hypotheses for thin plates, originally published in the mid-1800s, hence leading to the present day Love–Kirchhoff terminology.

For the transverse shear strains and stresses of the thick rotating cylindrical shells considered in Chapter 5, first-order shear deformation theory (FSDT) is employed, and these hypotheses thus do not apply for these cases.

In addition, unless otherwise stated, the present manuscript adopts the following scope and assumptions:

1.   The materials of the shells are linearly elastic.

2.   The shells possess constant thicknesses.

3.   Displacements vary linearly through the shell thickness.

7

4.   In the case of laminated composite shells, the layers are perfectly bonded together and each layer is of uniform thickness.

5.   The shells of revolution rotate about their longitudinal and symmetrical axis at a constant angular velocity.

2.2   Shell kinematic strain–displacement relations.

In shells, the geometric relationships between the strains and displacements, or kinematic relations, refer to the geometric changes or deformation, without consideration for the forces causing the deformation. Although structural shell systems may be nonlinear, over a certain geometrical range, the system can behave in a linear manner. In this restricted range, one can safely use linearized theories. This restriction is often called small-motion assumption, and the well-established acceptable range is

$$h/R_i \leq 0.05 \tag{2.1}$$

where h is the shell thickness and R_i refers to the smallest radius of curvature in the shell system under consideration.

For thin shell structures, there are various thin-shell theories and each is characterized by its own set of strain–displacement relations. The first mathematical shell theory was developed by Love [1888]. It is commonly known as Love's first approximation, and it was developed on the basis of Kirchhoff hypotheses for thin plate structures, first proposed in the mid-1800s. As such, Love's first approximation is also often referred to as the Love–Kirchhoff hypotheses for thin shells. These hypotheses form the foundation of all thin-shell theories. The originators of the different thin-shell theories have, at their discretion, defined the kinematic relations differently. In other words, the strain components, $\{\varepsilon_x, \varepsilon_\theta, \varepsilon_{x\theta}\}$, are defined by different expressions of the reference surface strains, $\{e_x, e_\theta, e_{x\theta}\}$, and reference surface curvatures, $\{\kappa_x, \kappa_\theta, \kappa_{x\theta}\}$, as well. As a result, the governing equations for the various thin-shell theories are also different from one another.

For thin cylindrical shells, the strain–displacement relations of several common thin-shell theories employed in this monograph are presented as follows.
Love's strain–displacement relations:

$$\varepsilon_x = e_x + z\kappa_x = \frac{\partial u}{\partial x} - z\frac{\partial^2 w}{\partial x^2} \tag{2.2}$$

$$\varepsilon_\theta = e_\theta + z\kappa_\theta = \frac{1}{R}\left(\frac{\partial v}{\partial \theta} + w\right) - \frac{z}{R^2}\left(\frac{\partial^2 w}{\partial \theta^2} - \frac{\partial v}{\partial \theta}\right) \tag{2.3}$$

$$\varepsilon_{x\theta} = e_{x\theta} + z\kappa_{x\theta} = \frac{\partial v}{\partial x} + \frac{1}{R}\frac{\partial u}{\partial \theta} - \frac{2z}{R}\left(\frac{\partial^2 w}{\partial x\,\partial \theta} - \frac{\partial v}{\partial x}\right) \tag{2.4}$$

where u, v and w are, respectively, the midsurface displacements in the meridional, x, circumferential, θ, and normal, z, directions of the cylindrical shells of revolution. Donnell's strain–displacement relations:

$$\varepsilon_x = e_x + z\kappa_x = \frac{\partial u}{\partial x} - z\frac{\partial^2 w}{\partial x^2} \tag{2.5}$$

$$\varepsilon_\theta = e_\theta + z\kappa_\theta = \frac{1}{R}\left(\frac{\partial v}{\partial \theta} + w\right) - \frac{z}{R^2}\frac{\partial^2 w}{\partial \theta^2} \tag{2.6}$$

$$\varepsilon_{x\theta} = e_{x\theta} + z\kappa_{x\theta} = \frac{\partial v}{\partial x} + \frac{1}{R}\frac{\partial u}{\partial \theta} - \frac{2z}{R}\frac{\partial^2 w}{\partial x\,\partial \theta}. \tag{2.7}$$

Flügge's strain–displacement relations:

$$\varepsilon_x = e_x + z\kappa_x = \frac{\partial u}{\partial x} - z\frac{\partial^2 w}{\partial x^2} \tag{2.8}$$

$$\varepsilon_\theta = e_\theta + z\kappa_\theta = \frac{1}{R}\frac{\partial v}{\partial \theta} + \frac{w}{R+z} - \frac{z}{(1+z/R)R^2}\frac{\partial^2 w}{\partial \theta^2} \tag{2.9}$$

$$\varepsilon_{x\theta} = e_{x\theta} + z\kappa_{x\theta} = \left(1 + \frac{z}{R}\right)\frac{\partial v}{\partial x} + \frac{1}{(1+z/R)R}\frac{\partial u}{\partial \theta} - \frac{z}{R}\left(1 + \frac{1}{1+z/R}\right)\frac{\partial^2 w}{\partial x\,\partial \theta}. \tag{2.10}$$

Sanders' strain–displacement relations:

$$\varepsilon_x = e_x + z\kappa_x = \frac{\partial u}{\partial x} - z\frac{\partial^2 w}{\partial x^2} \tag{2.11}$$

$$\varepsilon_\theta = e_\theta + z\kappa_\theta = \frac{1}{R}\left(\frac{\partial v}{\partial \theta} + w\right) - \frac{z}{R^2}\left(\frac{\partial^2 w}{\partial \theta^2} - \frac{\partial v}{\partial \theta}\right) \tag{2.12}$$

$$\varepsilon_{x\theta} = e_{x\theta} + z\kappa_{x\theta} = \frac{\partial v}{\partial x} + \frac{1}{R}\frac{\partial u}{\partial \theta} - z\left(\frac{1}{R}\left(2\frac{\partial^2 w}{\partial x\,\partial \theta} - \frac{3}{2}\frac{\partial v}{\partial x}\right) + \frac{1}{2R^2}\frac{\partial u}{\partial \theta}\right). \tag{2.13}$$

For thin truncated circular conical shells, the radius $r(x)$ is not a constant and varies with the meridional coordinate x

$$r(x) = a + x\sin\alpha \tag{2.14}$$

where α is the cone angle and a the radius at the small end of conical shell. The strain–displacement relations for the conical shell must thus include the effects of the cone angle α. The corresponding strain–displacement relations are written as

$$\varepsilon_x = e_x + z\kappa_x = \frac{\partial u}{\partial x} - z\frac{\partial^2 w}{\partial x^2} \tag{2.15}$$

$$\varepsilon_\theta = e_\theta + z\kappa_\theta = \frac{1}{r(x)}\frac{\partial v}{\partial \theta} + \frac{u\sin\alpha + w\cos\alpha}{r(x)}$$
$$- z\left(\frac{1}{r^2(x)}\frac{\partial^2 w}{\partial \theta^2} - \frac{\cos\alpha}{r^2(x)}\frac{\partial v}{\partial \theta} + \frac{\sin\alpha}{r(x)}\frac{\partial w}{\partial x}\right) \tag{2.16}$$

$$\varepsilon_{x\theta} = e_{x\theta} + z\kappa_{x\theta} = \frac{\partial v}{\partial x} + \frac{1}{r(x)}\frac{\partial u}{\partial \theta} - \frac{v\sin\alpha}{r(x)}$$
$$- 2z\left(\frac{1}{r(x)}\frac{\partial^2 w}{\partial x\,\partial \theta} - \frac{\sin\alpha}{r^2(x)}\frac{\partial w}{\partial \theta} - \frac{\cos\alpha}{r(x)}\frac{\partial v}{\partial x} + \frac{v\sin\alpha\cos\alpha}{r^2(x)}\right). \tag{2.17}$$

For the thick cylindrical shells, however, the Love–Kirchhoff hypothesis is relaxed by allowing the transverse normals to rotate about the midsurface during deformation. These rotations, about the θ and x axes, are respectively denoted by ψ_x and ψ_θ. Depending on the approximations made, the strain–displacement relations for thick cylindrical shells can also take various forms. For example, when the Reissner–Mindlin assumption is made and the well-known FSDT is employed, the transverse shear strains $\gamma^T = \{\varepsilon_{xz}, \varepsilon_{\theta z}\}$ are included in the strain–displacement relations

$$\varepsilon_x = e_x + z\kappa_x = \frac{\partial u}{\partial x} + z\frac{\partial \psi_x}{\partial x} \tag{2.18}$$

$$\varepsilon_\theta = e_\theta + z\kappa_\theta = \frac{1}{R}\frac{\partial v}{\partial \theta} + \frac{w}{R} + \frac{z}{R}\frac{\partial \psi_\theta}{\partial \theta} \tag{2.19}$$

$$\varepsilon_{x\theta} = e_{x\theta} + z\kappa_{x\theta} = \frac{\partial v}{\partial x} + \frac{1}{R}\frac{\partial u}{\partial \theta} + z\left(\frac{\partial \psi_\theta}{\partial x} + \frac{1}{R}\frac{\partial \psi_x}{\partial \theta} + \frac{1}{2R}\left(\frac{\partial v}{\partial x} - \frac{1}{R}\frac{\partial u}{\partial \theta}\right)\right) \tag{2.20}$$

$$\varepsilon_{xz} = \frac{\partial w}{\partial x} - \frac{v}{R} + \psi_x \tag{2.21}$$

$$\varepsilon_{\theta z} = \frac{1}{R}\frac{\partial w}{\partial \theta} + \psi_\theta. \tag{2.22}$$

Apart from the transverse shear strains $\gamma^T = \{\varepsilon_{xz}, \varepsilon_{\theta z}\}$, the above strain–displacement relations can all be expressed in a compact unified form as

$$\boldsymbol{\varepsilon} = \mathbf{e} + z\boldsymbol{\kappa} \tag{2.23}$$

where $\boldsymbol{\varepsilon}^T = \{\varepsilon_x, \varepsilon_\theta, \varepsilon_{x\theta}\}$ is the strain vector at a distance z from the reference surface (midsurface), $\mathbf{e}^T = \{e_x, e_\theta, e_{x\theta}\}$ the vector containing the reference surface strains and $\boldsymbol{\kappa}^T = \{\kappa_x, \kappa_\theta, \kappa_{x\theta}\}$ the vector containing the reference surface curvatures.

2.3   Resultant stress–strain relations in constitutive shell models.

Generally, constitutive shell models refer to the characteristics of the shells' material constitution and their response to applied loads. When linear elastic material is considered, the use of generic Hooke's law is valid for deriving the constitutive equations for general shells of revolution. Furthermore, for multi-layered shell laminates in which each layer is orthotropic, the stress–strain relation of the kth layer of the multi-layered shell laminate can be written as

$$\left\{\begin{array}{c}\sigma_x \\ \sigma_\theta \\ \sigma_{x\theta}\end{array}\right\}_k = \begin{bmatrix}\bar{Q}_{11} & \bar{Q}_{12} & \bar{Q}_{16} \\ \bar{Q}_{21} & \bar{Q}_{22} & \bar{Q}_{26} \\ \bar{Q}_{61} & \bar{Q}_{62} & \bar{Q}_{66}\end{bmatrix}_k \left\{\begin{array}{c}\varepsilon_x \\ \varepsilon_\theta \\ \varepsilon_{x\theta}\end{array}\right\}_k \tag{2.24}$$

or in standard vector form,

$$\boldsymbol{\sigma}_k = [\bar{Q}_{ij}]_k \boldsymbol{\varepsilon}_k = \bar{\mathbf{Q}}_k \boldsymbol{\varepsilon}_k \tag{2.25}$$

where $\sigma_k^T = \{\sigma_x, \sigma_\theta, \sigma_{x\theta}\}_k$ and $\varepsilon_k^T = \{\varepsilon_x, \varepsilon_\theta, \varepsilon_{x\theta}\}_k$ are the stress and strain vectors of the kth layer at a distance z from the reference surface of the shell, in which the subscripts x and θ represent, respectively, the meridional and circumferential directions. $\bar{\mathbf{Q}}_k = [\bar{Q}_{ij}]_k$ $(i,j = 1, 2, 6)$ is the transformed reduced stiffness matrix of the kth layer,

$$[\bar{Q}_{ij}]_k = [T_{ij}]^{-1}[Q_{ij}]_k[T_{ij}] \tag{2.26}$$

or in standard vector form,

$$\bar{\mathbf{Q}}_k = \mathbf{T}^{-1}\mathbf{Q}_k\mathbf{T} \tag{2.27}$$

where $\mathbf{T} = [T_{ij}]$ $(i,j = 1, 2, 3)$ is the transformation matrix between the material principal coordinates of the kth layer and the global geometric coordinates of the shell and it can be written as

$$\mathbf{T} = [T_{ij}] = \begin{bmatrix} \cos^2\phi & \sin^2\phi & 2\sin\phi\cos\phi \\ \sin^2\phi & \cos^2\phi & -2\sin\phi\cos\phi \\ -\sin\phi\cos\phi & \sin\phi\cos\phi & \cos^2\phi - \sin^2\phi \end{bmatrix} \tag{2.28}$$

where ϕ is the angle between the material principal direction of the kth layer and the global geometric coordinate direction of the rotating shell of revolution.

In Eqs. (2.26) and (2.27), $\mathbf{Q}_k = [Q_{ij}]_k$ is the reduced stiffness matrix of the kth layer. If the assumption of plane-stress state is made for the present rotating thin shell of revolution, $\mathbf{Q}_k = [Q_{ij}]_k$ for the kth orthotropic layer can be stated as follows:

$$\mathbf{Q}_k = [Q_{ij}]_k = \begin{bmatrix} \mu^* E_1^k & \mu^* \mu_{21} E_1^k & 0 \\ \mu^* \mu_{12} E_2^k & \mu^* E_2^k & 0 \\ 0 & 0 & G_{12}^k \end{bmatrix} \tag{2.29}$$

where $\mu^* = (1 - \mu_{12}^k \mu_{21}^k)^{-1}$, E_i^k, μ_{ij}^k and G_{ij}^k $(i,j = 1, 2)$ are the orthotropic elastic constants of the kth layer and $\mu_{21}^k E_1^k = \mu_{12}^k E_2^k$.

For the convenience of subsequent formulation of the governing equations of motion of rotating shells, Hooke's law is used in the form of resultant stress–strain relations for the entire multi-layered shell, which can be derived on basis of the above formulation of the kth layer. Therefore, one can define the stress resultants of the multi-layered shell, as shown in Figs. 2.1 and 2.2, $\mathbf{N}^T = \{N_x, N_\theta, N_{x\theta}\}$ normally termed the

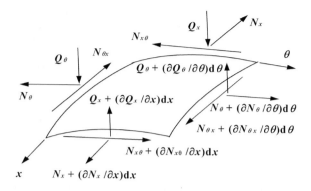

Figure 2.1 Notation and positive directions of force
resultants in shell coordinates.

in-plane force resultants, and $\mathbf{M}^{T} = \{M_x, M_\theta, M_{x\theta}\}$ normally termed the moment
resultants, as follows

$$\mathbf{N}^{T} = \{N_x, N_\theta, N_{x\theta}\} = \int_{-h/2}^{h/2} \{\sigma_x, \sigma_\theta, \sigma_{x\theta}\}dz \qquad (2.30)$$

$$\mathbf{M}^{T} = \{M_x, M_\theta, M_{x\theta}\} = \int_{-h/2}^{h/2} \{\sigma_x, \sigma_\theta, \sigma_{x\theta}\}z \, dz \qquad (2.31)$$

where $N_{x\theta} = N_{\theta x}$ and $M_{x\theta} = M_{\theta x}$, h is the thickness of the shell, $\boldsymbol{\sigma}^{T} = \{\sigma_x, \sigma_\theta, \sigma_{x\theta}\}$ the
stress vector in which σ_x, σ_θ and $\sigma_{x\theta}$ are the meridional, circumferential and shear
stresses, respectively. In particular, when Flügge's thin-shell theory is implemented for

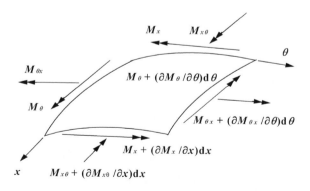

Figure 2.2 Notation and positive directions of
moment resultants in shell coordinates.

rotating cylindrical shells

$$\mathbf{N}^{\mathrm{T}} = \{N_x, N_\theta, N_{x\theta}, N_{\theta x}\} = \int_{-h/2}^{h/2} \{\sigma_x(1 + z/R), \sigma_\theta, \sigma_{x\theta}(1 + z/R), \sigma_{\theta x}\}\mathrm{d}z \qquad (2.32)$$

$$\mathbf{M}^{\mathrm{T}} = \{M_x, M_\theta, M_{x\theta}, M_{\theta x}\} = \int_{-h/2}^{h/2} \{\sigma_x(1 + z/R), \sigma_\theta, \sigma_{x\theta}(1 + z/R), \sigma_{\theta x}\}z\,\mathrm{d}z. \qquad (2.33)$$

Substituting Eqs. (2.23) and (2.26) together with Eqs. (2.28) and (2.29) into Eq. (2.24), followed by further substitution of the resulting expressions into Eqs. (2.30) and (2.31) or Eqs. (2.32) and (2.33), the general constitutive relationship of the multi-layered shell is derived as

$$\begin{Bmatrix} \mathbf{N} \\ \mathbf{M} \end{Bmatrix} = \begin{bmatrix} \mathbf{A} & \mathbf{B} \\ \mathbf{B} & \mathbf{D} \end{bmatrix} \begin{Bmatrix} \mathbf{e} \\ \mathbf{\kappa} \end{Bmatrix} \qquad (2.34)$$

where $\mathbf{A} = [A_{ij}]$, $\mathbf{B} = [B_{ij}]$ and $\mathbf{D} = [D_{ij}]$ $(i, j = 1, 2, 6)$ are, respectively, the tensile, coupling and bending stiffness matrices, and these are defined as

$$(A_{ij}, B_{ij}, D_{ij}) = \int_{-h/2}^{h/2} \bar{Q}_{ij}(1, z, z^2)\mathrm{d}z. \qquad (2.35)$$

When the multi-layered shell is reduced to an orthotropic single-layer shell, the tensile, bending and coupling stiffnesses are, respectively, reduced to

$$A_{11} = \frac{E_x h}{1 - \mu_{x\theta}\mu_{\theta x}}, \qquad A_{22} = \frac{E_\theta h}{1 - \mu_{x\theta}\mu_{\theta x}}, \qquad A_{12} = \frac{\mu_{\theta x}E_x h}{1 - \mu_{x\theta}\mu_{\theta x}},$$

$$A_{21} = \frac{\mu_{x\theta}E_\theta h}{1 - \mu_{x\theta}\mu_{\theta x}}, \qquad A_{66} = G_{x\theta}, \qquad A_{16} = A_{61} = A_{26} = A_{62} = 0,$$

$$D_{16} = D_{61} = D_{26} = D_{62} = 0, \qquad B_{ij} = 0 \ (i, j = 1, 2, 6),$$

$$D_{11} = \frac{E_x h^3}{12(1 - \mu_{x\theta}\mu_{\theta x})}, \qquad D_{22} = \frac{E_\theta h^3}{12(1 - \mu_{x\theta}\mu_{\theta x})},$$

$$D_{12} = \frac{\mu_{\theta x}E_x h^3}{12(1 - \mu_{x\theta}\mu_{\theta x})}, \qquad D_{21} = \frac{\mu_{x\theta}E_\theta h^3}{12(1 - \mu_{x\theta}\mu_{\theta x})}, \qquad D_{66} = \frac{G_{x\theta}h^3}{12} \qquad (2.36)$$

in which E_x, E_θ, $\mu_{x\theta}$, $\mu_{\theta x}$ and $G_{x\theta}$ are orthotropic elastic constants and $\mu_{x\theta}E_\theta = \mu_{\theta x}E_x$.

Further simplification to an isotropic single-layered shell, the tensile, bending and coupling stiffnesses are thus further reduced to

$$A_{11} = A_{22} = \frac{Eh}{(1 - \mu^2)}, \qquad A_{12} = A_{21} = \frac{\mu Eh}{(1 - \mu^2)}, \qquad A_{66} = \frac{Eh}{2(1 + \mu)},$$

$$A_{16} = A_{61} = A_{26} = A_{62} = 0, \qquad D_{11} = D_{22} = \frac{Eh^3}{12(1 - \mu^2)},$$

$$D_{12} = D_{21} = \frac{\mu Eh^3}{12(1 - \mu^2)}, \qquad D_{66} = \frac{Eh^3}{12(1 + \mu)},$$

$$D_{16} = D_{61} = D_{26} = D_{62} = 0, \qquad B_{ij} = 0 \; (i, j = 1, 2, 6) \tag{2.37}$$

where E and μ are Young's modulus and Poisson's ratio, respectively.

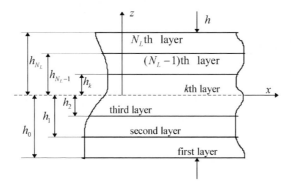

Figure 2.3 Cross-sectional view of the multi-layered shell.

For the general case of an arbitrarily laminated composite shell, as shown in Fig. 2.3, the tensile, coupling and bending stiffnesses are

$$(A_{ij}, B_{ij}, D_{ij}) = \int_{-h/2}^{h/2} \bar{Q}_{ij}(1, z, z^2) dz = \sum_{k=1}^{N_L} \int_{z_k}^{z_{k+1}} \bar{Q}_{ij}^{(k)}(1, z, z^2) dz \tag{2.38}$$

which after carrying out the simple integrations can be rewritten as

$$A_{ij} = \sum_{k=1}^{N_L} \bar{Q}_{ij}^{(k)}(h_k - h_{k-1}), \qquad B_{ij} = \frac{1}{2} \sum_{k=1}^{N_L} \bar{Q}_{ij}^{(k)}(h_k^2 - h_{k-1}^2),$$

$$\tag{2.39}$$

$$D_{ij} = \frac{1}{3} \sum_{k=1}^{N_L} \bar{Q}_{ij}^{(k)}(h_k^3 - h_{k-1}^3)$$

where $\bar{Q}_{ij}^{(k)}$ represents the elements of the transformed reduced stiffness matrix $\bar{\mathbf{Q}}_k$, see Eq. (2.26), for the kth layer. h_k and h_{k-1}, respectively, denote the distances of the outer and inner surfaces of the kth layer from the reference surface (see Fig. 2.3). N_L is the total number of layers of the laminated composite shell. For example, if we take $N_L = 3$, the various stiffnesses (A_{ij}, B_{ij}, D_{ij}) will therefore be obtained for a sandwich-type shell as shown in Fig. 2.4.

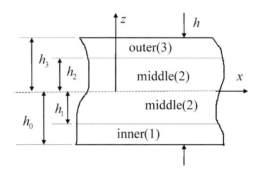

Figure 2.4 Cross-sectional view of thickness
of the sandwich-type shell.

Based on the above formulations, one can express the governing equations of motion in terms of the displacements for the dynamic analysis of the rotating shells of revolution.

2.4   Governing equations of motion.

In this section, we establish the fundamental governing equations of motion for a general rotating shell of revolution. By transformation of the curvilinear coordinates, the dynamic governing equations in terms of the forces, moments and displacements can then be reduced and derived for specific shells of revolution, such as the rotating cylindrical shells in Chapters 3 and 6, and rotating conical shells in Chapter 4.

According to Flügge's shell theory [1960], a nonlinear shell problem may be approximated as two linear problems in which one may use linear shell theories. In the present analysis, this concept of linear approximation is adopted, and the derivatives of the vector bases and Lamé parameters are used. Based on the formulative work by Chen *et al.* [1993], the fundamental dynamic equilibrium equations are constructed for rotating shells of revolution. The resulting set of equilibrium governing equations is subsequently expanded asymptotically into two groups of equations, with reference to the basic and the additional states, respectively, where linear shell theories can then be employed.

Consider a thin shell of revolution rotating about its longitudinal and symmetrical axis at a constant angular velocity Ω as shown in Fig. 2.5. A curvilinear coordinate system is also shown, in which $\mathbf{s}_1$, $\mathbf{s}_2$ and $\mathbf{s}_n$ are the unit vectors of the curvilinear coordinate axes. The subscript n denotes the normal direction, while α_1 and α_2 represent meridian and parallel coordinates, respectively. From Fig. 2.5, it is observed that, based on the coordinate system of the shell of revolution (φ, θ, z), the positions of the points on the middle surface of the shell of arbitrary shape can be determined by the angles, θ and φ. R_1 is the radius of curvature of the meridian, and R_2 is the radius of curvature of the normal section, tangential to the horizontal planar circle. The latter radius R_2 is also equal to the segment perpendicular to the middle surface, taken from the point of this surface to

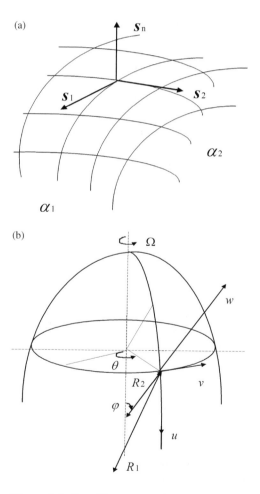

Figure 2.5 Curvilinear coordinate system of
a shell of revolution.

the longitudinal symmetrical axis of the shell. Obviously both these radii are functions of the angle φ. In the following formulations, the fundamental governing equations of motion are established for the states of deformation of the respective shell elements. The subscript "d" denotes the variables which have been deformed.

According to the conservation of mass in the infinitesimal body before and after deformation, we have

$$(\rho h)_d = \rho h / (1 + e_1 + e_2) \approx \rho h (1 - e_1 - e_2) \tag{2.40}$$

where ρ and h are, respectively, the mass density and thickness of shell, and e_1 and e_2 are the normal strains.

The angle in the α_1 direction of principal coordinate axis after deformation is

$$(\varphi)_d = \varphi + \vartheta \tag{2.41}$$

with (ϑ is a rotational angle normal to α_1)

$$\sin(\varphi)_d \approx \sin \varphi + \vartheta \cos \varphi \tag{2.42}$$

and the radius thus becomes

$$(r)_d \approx r(1 + e_2). \tag{2.43}$$

Based on the Novozhilov shell theory [1964], the relationships between the Lamé parameters before and after deformations are given as follows

$$A_{1d} \approx A_1(1 + e_1) \qquad \text{and} \qquad A_{2d} \approx A_2(1 + e_2) \tag{2.44}$$

and the relationships between the vector bases before and after deformations are written as

$$\mathbf{s}_{1d} \approx \mathbf{s}_1 + \beta_1 \mathbf{s}_2 - \vartheta \mathbf{s}_n, \qquad \mathbf{s}_{2d} \approx \mathbf{s}_2 + \beta_2 \mathbf{s}_1 - \psi \mathbf{s}_n, \qquad \mathbf{s}_{nd} \approx \mathbf{s}_n + \vartheta \mathbf{s}_1 + \psi \mathbf{s}_2 \tag{2.45}$$

in which ψ is a rotational angle normal to α_2, while the strain–displacement relationships of the shells of revolution can be expressed as

$$e_1 = \frac{1}{A_1} \frac{\partial u}{\partial \alpha_1} + \frac{1}{A_1 A_2} \frac{\partial A_1}{\partial \alpha_2} v + \frac{w}{R_1}, \qquad e_2 = \frac{1}{A_2} \frac{\partial v}{\partial \alpha_2} + \frac{1}{A_1 A_2} \frac{\partial A_2}{\partial \alpha_1} u + \frac{w}{R_2} \tag{2.46}$$

$$\beta_1 = \frac{1}{A_1}\frac{\partial v}{\partial \alpha_1} - \frac{1}{A_1 A_2}\frac{\partial A_1}{\partial \alpha_2}u, \qquad \beta_2 = \frac{1}{A_2}\frac{\partial u}{\partial \alpha_2} - \frac{1}{A_1 A_2}\frac{\partial A_2}{\partial \alpha_1}v \qquad (2.47)$$

$$\vartheta = -\frac{1}{A_1}\frac{\partial w}{\partial \alpha_1} + \frac{u}{R_1}, \qquad \psi = -\frac{1}{A_2}\frac{\partial w}{\partial \alpha_2} + \frac{v}{R_2}. \qquad (2.48)$$

The derivatives of the vector bases are formulated as

$$\frac{\partial \mathbf{s}_1}{\partial \alpha_1} = -\frac{1}{A_2}\frac{\partial A_1}{\partial \alpha_2}\mathbf{s}_2 - \frac{A_1}{R_1}\mathbf{s}_n, \qquad \frac{\partial \mathbf{s}_2}{\partial \alpha_2} = -\frac{1}{A_1}\frac{\partial A_2}{\partial \alpha_1}\mathbf{s}_1 - \frac{A_2}{R_2}\mathbf{s}_n \qquad (2.49)$$

$$\frac{\partial \mathbf{s}_1}{\partial \alpha_2} = \frac{1}{A_1}\frac{\partial A_2}{\partial \alpha_1}\mathbf{s}_2, \qquad \frac{\partial \mathbf{s}_2}{\partial \alpha_1} = -\frac{1}{A_2}\frac{\partial A_1}{\partial \alpha_2}\mathbf{s}_1 \qquad (2.50)$$

$$\frac{\partial \mathbf{s}_n}{\partial \alpha_1} = \frac{A_1}{R_1}\mathbf{s}_1, \qquad \frac{\partial \mathbf{s}_n}{\partial \alpha_2} = \frac{A_2}{R_2}\mathbf{s}_2. \qquad (2.51)$$

By the above relationships of the Novozhilov shell theory [1964], the forces and velocities for the infinitesimal body can be stated, respectively, as follows:

(1) Centripetal force

$$\mathbf{q}_1 = (\rho h)_d \Omega^2 (r)_d [(\cos \varphi)\mathbf{s}_{1d} + (\sin \varphi)\mathbf{s}_{nd}]$$

$$\approx \rho h \Omega^2 r[(\cos \varphi + \vartheta \sin \varphi - e_1 \cos \varphi)\mathbf{s}_1 + (\beta_1 \cos \varphi + \psi \sin \varphi)\mathbf{s}_2$$

$$+ (\sin \varphi - \vartheta \cos \varphi - e_1 \sin \varphi)\mathbf{s}_n]. \qquad (2.52)$$

(2) Inertia force

$$\mathbf{q}_2 = -\rho h \frac{\partial^2 u}{\partial t^2}\mathbf{s}_{1d} - \rho h \frac{\partial^2 v}{\partial t^2}\mathbf{s}_{2d} - \rho h \frac{\partial^2 w}{\partial t^2}\mathbf{s}_{nd}$$

$$\approx -\rho h \frac{\partial^2 u}{\partial t^2}\mathbf{s}_1 - \rho h \frac{\partial^2 v}{\partial t^2}\mathbf{s}_2 - \rho h \frac{\partial^2 w}{\partial t^2}\mathbf{s}_n. \qquad (2.53)$$

(3) Angular velocity

$$\mathbf{a} = \Omega[(\sin \varphi)\mathbf{s}_{1d} + (\cos \varphi)\mathbf{s}_{nd}]. \qquad (2.54)$$

(4)   Relative velocity

$$\mathbf{v}_r = \frac{\partial u}{\partial t}\mathbf{s}_{1d} + \frac{\partial v}{\partial t}\mathbf{s}_{2d} + \frac{\partial w}{\partial t}\mathbf{s}_{nd}. \tag{2.55}$$

(5)   Coriolis force

$$\mathbf{q}_3 = -2(\rho h)_d \mathbf{a} \times \mathbf{v}_r$$

$$= 2\rho h \Omega \left[\left(\frac{\partial v}{\partial t}\cos\varphi \right)\mathbf{s}_{1d} - \left(\frac{\partial u}{\partial t}\cos\varphi + \frac{\partial w}{\partial t}\sin\varphi \right)\mathbf{s}_{2d} + \left(\frac{\partial v}{\partial t}\sin\varphi \right)\mathbf{s}_{nd} \right]$$

$$\approx 2\rho h \Omega \left[\left(\frac{\partial v}{\partial t}\cos\varphi \right)\mathbf{s}_1 - \left(\frac{\partial u}{\partial t}\cos\varphi + \frac{\partial w}{\partial t}\sin\varphi \right)\mathbf{s}_2 + \left(\frac{\partial v}{\partial t}\sin\varphi \right)\mathbf{s}_n \right] \tag{2.56}$$

in which $\times$ denotes the vector product.

(6)   Inner forces

$$\mathbf{N}^{(1)} = (N_1\mathbf{s}_{1d} + N_{12}\mathbf{s}_{2d} + Q_1\mathbf{s}_{nd})A_{2d}d\alpha_2$$

$$\approx [(N_1 + N_{12}\beta_2 + Q_1\vartheta)\mathbf{s}_1 + (N_1\beta_1 + N_{12} + Q_1\psi)\mathbf{s}_2$$

$$+ (-N_1\vartheta - N_{12}\psi + Q_1)\mathbf{s}_n]A_{2d}d\alpha_2$$

$$\mathbf{N}^{(2)} = (N_{21}\mathbf{s}_{1d} + N_2\mathbf{s}_{2d} + Q_2\mathbf{s}_{nd})A_{1d}d\alpha_1$$

$$\approx [(N_{21} + N_2\beta_2 + Q_2\vartheta)\mathbf{s}_1 + (N_{21}\beta_1 + N_2 + Q_2\psi)\mathbf{s}_2$$

$$+ (-N_{21}\vartheta - N_2\psi + Q_2)\mathbf{s}_n]A_{1d}d\alpha_1. \tag{2.57}$$

(7)   Inner moments

$$\mathbf{M}^{(1)} = (M_1\mathbf{s}_{2d} - M_{12}\mathbf{s}_{1d})A_{2d}d\alpha_2$$

$$\approx [(M_1\beta_2 - M_{12})\mathbf{s}_1 + (M_1 - M_{12}\beta_1)\mathbf{s}_2 + (-M_1\psi + M_{12}\vartheta)\mathbf{s}_n]A_{2d}d\alpha_2$$

$$\mathbf{M}^{(2)} = (-M_2\mathbf{s}_{1d} + M_{21}\mathbf{s}_{2d})A_{1d}d\alpha_1$$

$$\approx [(-M_2 + M_{21}\beta_1)\mathbf{s}_1 + (-M_2\beta_2 + M_{21})\mathbf{s}_2 + (M_2\vartheta - M_{21}\psi)\mathbf{s}_n]A_{1d}d\alpha_1. \tag{2.58}$$

Therefore, the dynamic equilibrium equations can be developed in the following form

$$hskip - 23pt \frac{\partial \mathbf{N}^{(1)}}{\partial \alpha_1} d\alpha_1 + \frac{\partial \mathbf{N}^{(2)}}{\partial \alpha_2} d\alpha_2 + \sum_i \mathbf{q}_i A_{1d} A_{2d} \, d\alpha_1 \, d\alpha_2 = 0$$

$$\frac{\partial \mathbf{M}^{(1)}}{\partial \alpha_1} d\alpha_1 + \frac{\partial \mathbf{M}^{(2)}}{\partial \alpha_2} d\alpha_2 + \sum_i \mathbf{M}_i A_{1d} A_{2d} \, d\alpha_1 \, d\alpha_2 = 0$$

(2.59)

where $\sum_i \mathbf{M}_i$ is a sum of the moments resulting from Q_1 and Q_2, as well as other forces due to nonlinear effects, namely

$$\sum_i \mathbf{M}_i = [Q_2 - Q_1 \beta_2 + (N_{12} - N_{21})\vartheta] \mathbf{s}_1 - [-Q_2 \beta_1 + Q_1 - (N_{12} - N_{21})\psi] \mathbf{s}_2$$

$$- [-Q_2 \vartheta + Q_1 \psi + (N_{12} - N_{21})] \mathbf{s}_n.$$

(2.60)

On the basis of the linear approximations of Flügge's shell theory [1960], the resultant forces and moments (see Fig. 2.6) can be considered as linear combinations of the basic and additional states, namely

$$N_{ij} = N_{Bij} + N_{Aij} \qquad \text{and} \qquad M_{ij} = M_{Bij} + M_{Aij}$$

(2.61)

where subscript B denotes the basic state variations and subscript A the additional state variations.

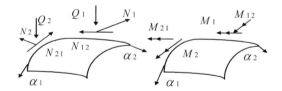

Figure 2.6 Directions of the force and moment resultants.

Substituting the relations of linear combinations (2.61) into the dynamic equilibrium equations (2.59), followed by asymptotically expanding the resulting expressions into the two groups of equations, the first- and second-order terms of the expanded equations can thus be collected for the basic and additional states, respectively.

For the basic state, the derived equations constitute the first-order terms of the asymptotically expanded equations for the equilibrium description of the centripetal forces. They are expressed as follows

$$\frac{1}{A_1A_2}\left[\frac{\partial A_2 N_{B1}}{\partial \alpha_1} + \frac{\partial A_1 N_{B21}}{\partial \alpha_2} + \frac{\partial A_1}{\partial \alpha_2}N_{B12} - \frac{\partial A_2}{\partial \alpha_1}N_{B2}\right]$$
$$+ \frac{Q_{B1}}{R_1} + \rho h \Omega^2 r \cos \varphi = 0 \tag{2.62}$$

$$\frac{1}{A_1A_2}\left[\frac{\partial A_2 N_{B12}}{\partial \alpha_1} + \frac{\partial A_1 N_{B2}}{\partial \alpha_2} + \frac{\partial A_2}{\partial \alpha_1}N_{B21} - \frac{\partial A_1}{\partial \alpha_2}N_{B1}\right] + \frac{Q_{B2}}{R_2} = 0 \tag{2.63}$$

$$\frac{1}{A_1A_2}\left[\frac{\partial A_2 Q_{B1}}{\partial \alpha_1} + \frac{\partial A_1 Q_{B2}}{\partial \alpha_2}\right] - \frac{N_{B1}}{R_1} - \frac{N_{B2}}{R_2} + \rho h \Omega^2 r \sin \varphi = 0 \tag{2.64}$$

$$\frac{1}{A_1A_2}\left[\frac{\partial A_2 M_{B1}}{\partial \alpha_1} + \frac{\partial A_1 M_{B21}}{\partial \alpha_2} + \frac{\partial A_1}{\partial \alpha_2}M_{B12} - \frac{\partial A_2}{\partial \alpha_1}M_{B2}\right] - Q_{B1} = 0 \tag{2.65}$$

$$\frac{1}{A_1A_2}\left[\frac{\partial A_2 M_{B12}}{\partial \alpha_1} + \frac{\partial A_1 M_{B2}}{\partial \alpha_2} + \frac{\partial A_2}{\partial \alpha_1}M_{B21} - \frac{\partial A_1}{\partial \alpha_2}M_{B1}\right] - Q_{B2} = 0 \tag{2.66}$$

$$\frac{M_{B12}}{R_1} - \frac{M_{B21}}{R_2} + N_{B12} - N_{B21} = 0. \tag{2.67}$$

For the additional state, the formulated equations are based on the second-order terms of the asymptotically expanded equations for the motional description with consideration of the Coriolis forces and nonlinear effects due to the large centripetal force. They are written as

$$\frac{1}{A_1A_2}\left[\frac{\partial A_2 N_{A1}}{\partial \alpha_1} + \frac{\partial A_1 N_{A21}}{\partial \alpha_2} + \frac{\partial A_1}{\partial \alpha_2}N_{A12} - \frac{\partial A_2}{\partial \alpha_1}N_{A2}\right] + \frac{Q_{A1}}{R_1} + \frac{\partial e_2}{A_1 \partial \alpha_1}N_{B1}$$

$$+ \frac{\partial e_1}{A_2 \partial \alpha_2}N_{B21} + \frac{1}{A_1A_2}\left[-\frac{\partial A_2 N_{B1}}{\partial \alpha_1}e_1 - \frac{\partial A_1 N_{B21}}{\partial \alpha_2}e_2 - \frac{\partial A_1}{\partial \alpha_2}N_{B12}e_2\right.$$

$$+ \frac{\partial A_2}{\partial \alpha_1}N_{B2}e_1 + \frac{\partial A_1}{\partial \alpha_2}N_{B1}\beta_1 - \frac{\partial A_2}{\partial \alpha_1}N_{B21}\beta_1 + \frac{\partial A_1 N_{B2}\beta_2}{\partial \alpha_2} + \frac{\partial A_2 N_{B12}\beta_2}{\partial \alpha_1}$$

$$+ \frac{\partial A_2 Q_{B1}\vartheta}{\partial \alpha_1} + \frac{\partial A_1 Q_{B2}\vartheta}{\partial \alpha_2} + \frac{\partial A_1}{\partial \alpha_2}Q_{B1}\psi - \frac{\partial A_2}{\partial \alpha_1}Q_{B2}\psi\right] - \frac{N_{B1}\vartheta}{R_1}$$

$$- \frac{N_{B12}\psi}{R_1} - \frac{Q_{B1}e_1}{R_1} - \rho h \Omega^2 r e_1 \cos \varphi + \rho h \Omega^2 r \vartheta \sin \varphi$$

$$+ 2\rho h \Omega \frac{\partial v}{\partial t}\cos \varphi - \rho h \frac{\partial^2 u}{\partial t^2} = 0 \tag{2.68}$$

$$\frac{1}{A_1 A_2}\left[\frac{\partial A_2 N_{A12}}{\partial \alpha_1} + \frac{\partial A_1 N_{A2}}{\partial \alpha_2} + \frac{\partial A_2}{\partial \alpha_1} N_{A21} - \frac{\partial A_1}{\partial \alpha_2} N_{A1}\right] + \frac{Q_{A2}}{R_2} + \frac{\partial e_2}{A_1 \partial \alpha_1} N_{B12}$$

$$+ \frac{\partial e_1}{A_2 \partial \alpha_2} N_{B2} + \frac{1}{A_1 A_2}\left[-\frac{\partial A_2 N_{B12}}{\partial \alpha_1} e_1 - \frac{\partial A_1 N_{B2}}{\partial \alpha_2} e_2 - \frac{\partial A_2}{\partial \alpha_1} N_{B21} e_1\right.$$

$$+ \frac{\partial A_1}{\partial \alpha_2} N_{B1} e_2 - \frac{\partial A_1}{\partial \alpha_2} N_{B12}\beta_2 + \frac{\partial A_2}{\partial \alpha_1} N_{B2}\beta_2 + \frac{\partial A_1 N_{B21}\beta_1}{\partial \alpha_2} + \frac{\partial A_2 N_{B1}\beta_1}{\partial \alpha_1}$$

$$\left. + \frac{\partial A_2 Q_{B1}\psi}{\partial \alpha_1} + \frac{\partial A_1 Q_{B2}\psi}{\partial \alpha_2} - \frac{\partial A_1}{\partial \alpha_2} Q_{B1}\vartheta + \frac{\partial A_2}{\partial \alpha_1} Q_{B2}\vartheta\right] - \frac{N_{B21}\vartheta}{R_2} - \frac{N_{B2}\psi}{R_2}$$

$$- \frac{Q_{B2} e_2}{R_2} + \rho h \Omega^2 r(\psi \sin\varphi + \beta_1 \cos\varphi)$$

$$- 2\rho h \Omega\left(\frac{\partial u}{\partial t}\cos\varphi + \frac{\partial w}{\partial t}\sin\varphi\right) - \rho h \frac{\partial^2 v}{\partial t^2} = 0 \qquad (2.69)$$

$$\frac{1}{A_1 A_2}\left[\frac{\partial A_2 Q_{A1}}{\partial \alpha_1} + \frac{\partial A_1 Q_{A2}}{\partial \alpha_2}\right] - \frac{N_{A1}}{R_1} - \frac{N_{A2}}{R_2}$$

$$+ \frac{1}{A_1 A_2}\left[-\frac{\partial A_2 N_{B1}\vartheta}{\partial \alpha_1} - \frac{\partial A_1 N_{B21}\vartheta}{\partial \alpha_2} - \frac{\partial A_2 N_{B12}\psi}{\partial \alpha_1} - \frac{\partial A_1 N_{B2}\psi}{\partial \alpha_2}\right.$$

$$\left. - \frac{\partial A_2 Q_{B1}}{\partial \alpha_1} e_1 - \frac{\partial A_1 Q_{B2}}{\partial \alpha_2} e_2\right] + \frac{1}{R_1}(-N_{B12}\beta_2 + N_{B1} e_1 - Q_{B1}\vartheta)$$

$$+ \frac{1}{R_2}(-N_{B21}\beta_1 + N_{B2} e_2 - Q_{12}\psi) - \rho h \Omega^2 r e_1 \sin\varphi$$

$$+ 2\rho h \Omega \frac{\partial w}{\partial t}\sin\varphi - \rho h \frac{\partial^2 w}{\partial t^2} = 0 \qquad (2.70)$$

$$\frac{1}{A_1 A_2}\left[\frac{\partial A_2 M_{A1}}{\partial \alpha_1} + \frac{\partial A_1 M_{A21}}{\partial \alpha_2} + \frac{\partial A_1}{\partial \alpha_2} M_{A12} - \frac{\partial A_2}{\partial \alpha_1} M_{A2}\right]$$

$$- Q_{A1} + \frac{\partial e_2}{A_1 \partial \alpha_1} M_{B1} + \frac{\partial e_1}{A_2 \partial \alpha_2} M_{B21} + \frac{1}{A_1 A_2}\left[-\frac{\partial A_2 M_{B1}}{\partial \alpha_1} e_1\right.$$

$$- \frac{\partial A_1 M_{B21}}{\partial \alpha_2} e_2 - \frac{\partial A_1}{\partial \alpha_2} M_{B12} e_2 + \frac{\partial A_2}{\partial \alpha_1} M_{B2} e_1 - \frac{\partial A_1}{\partial \alpha_2} M_{B1}\beta_2$$

$$\left. + \frac{\partial A_2}{\partial \alpha_1} M_{B21}\beta_2 - \frac{\partial A_1 M_{B2}\beta_1}{\partial \alpha_2} - \frac{\partial A_2 M_{B12}\beta_1}{\partial \alpha_1}\right] + \frac{M_{B2}\vartheta}{R_2}$$

$$- \frac{M_{B21}\psi}{R_2} + Q_{B2}\beta_1 + (N_{B12} - N_{B21})\psi = 0 \qquad (2.71)$$

$$\frac{1}{A_1 A_2}\left[\frac{\partial A_2 M_{A12}}{\partial \alpha_1} + \frac{\partial A_1 M_{A2}}{\partial \alpha_2} + \frac{\partial A_2}{\partial \alpha_1}M_{A21} - \frac{\partial A_1}{\partial \alpha_2}M_{A1}\right] - Q_{A2} + \frac{\partial e_2}{A_1 \partial \alpha_1}M_{B12}$$

$$+ \frac{\partial e_1}{A_2 \partial \alpha_2}M_{B2} + \frac{1}{A_1 A_2}\left[-\frac{\partial A_2 M_{B12}}{\partial \alpha_1}e_1 - \frac{\partial A_1 M_{B2}}{\partial \alpha_2}e_2 - \frac{\partial A_2}{\partial \alpha_1}M_{B21}e_1 + \frac{\partial A_1}{\partial \alpha_2}M_{B1}e_2\right.$$

$$\left. + \frac{\partial A_1}{\partial \alpha_2}M_{B12}\beta_1 - \frac{\partial A_2}{\partial \alpha_1}M_{B2}\beta_1 - \frac{\partial A_1 M_{B21}\beta_2}{\partial \alpha_2} - \frac{\partial A_2 M_{B1}\beta_2}{\partial \alpha_1}\right]$$

$$+ \frac{M_{B1}\psi}{R_1} - \frac{M_{B12}\vartheta}{R_1} - Q_{B1}\beta_2 + (N_{B12} - N_{B21})\psi = 0 \qquad (2.72)$$

$$\frac{M_{A12}}{R_1} - \frac{M_{A21}}{R_2} + N_{A12} - N_{A21} - \frac{M_{B1}\beta_2}{R_1} + \frac{M_{B2}\beta_1}{R_2} + \frac{M_{B12}\beta_1}{R_1} + \frac{M_{B21}\beta_2}{R_2}$$

$$+ \frac{1}{A_1 A_2}\left[-\frac{\partial A_2 M_{B12}\vartheta}{\partial \alpha_1} - \frac{\partial A_1 M_{B2}\vartheta}{\partial \alpha_2} + \frac{\partial A_2 M_{B1}\psi}{\partial \alpha_1} - \frac{\partial A_1 M_{B21}\psi}{\partial \alpha_2}\right] = 0. \qquad (2.73)$$

The above two groups of derived equations, Eqs. (2.62)–(2.67) and Eqs. (2.68)–(2.73), form the fundamental governing equations for rotating shells of revolution. Based on these equations, by the transformations of curvilinear coordinates and the adoption of various hypotheses, assumptions and restrictions described in Section 2.1, the governing equations in terms of forces, moments and displacements can be reduced for the rotating cylindrical and conical shells to be considered in the following chapters.

2.5   Eigenvalue analysis of boundary value problems.

From the mathematical viewpoint, a boundary value problem is a typical one described by ordinary differential equations or partial differential equations, and which have assigned values on the physical boundaries of the domain under consideration. All the problems studied in this monograph fall into this category. These are defined by a set of governing equations of motion consisting of partial differential equations, and several constrained boundary conditions at the edges of a rotating shell of revolution. For the definition of eigenvalue, it is always with reference to a square matrix. Namely if $\mathbf{A}$ is an $n \times n$ matrix, an eigenvalue of this characteristic matrix $\mathbf{A}$ is defined as a scalar λ such that the vector equation $\mathbf{A}\mathbf{x} = \lambda\mathbf{x}$ holds for a certain nonzero vector $\mathbf{x}$. Then λ is said to be an eigenvalue or characteristic value of the matrix $\mathbf{A}$, and the nonzero $\mathbf{x}$ the corresponding eigenvector or characteristic vector.

From the standpoint of engineering application, typical solutions for global structural characteristics such as natural frequencies of free vibration and critical buckling loads invariably involve eigenvalue analysis, and these are among the most important problems in connection with matrices. The number of research papers on various numerical methods to obtain the characteristic matrices for associated static and dynamic structural problems is enormous.

In this monograph, the boundary value problems will be derived for the dynamics of rotating shell structures of revolution. These will then be reduced via various numerical methods to standard eigenvalue problems. Specific problems of interest here include the free vibration, critical speed and dynamic stability analyses of rotating cylindrical and conical shells. The reader will find a host of information with regard to the influences of material and geometrical parameters on the different deformation modes of the various dynamic analyses carried out. Primarily, these studies will meet the needs of scientists and engineers in the broad areas of aeronautics, astronautics, and civil and mechanical engineering, being especially useful as a reference source in the computational modeling and simulation aspects of rotating shell-like components. Furthermore, they will also be invaluable resource to application design engineers in aeronautical and aerospace industries, with extensive benchmark results to compare and verify their experimental data against.

Chapter 3
Free Vibration of Thin Rotating Cylindrical Shells

3.1   Introduction.

Since the first recorded study on the vibration of a rotating ring (as a special case of rotating shell of revolution) was carried out and the traveling-wave phenomenon was first discovered by Bryan [1890], there have been substantial published works on the dynamics of the rotating cylindrical shells. Early works examined the effects of Coriolis forces and this was first discussed by DiTaranto & Lessen [1964]. The coupled effects of the Coriolis forces and traveling waves were initially analyzed by Srinivasan & Lauterbach [1971] for an infinitely long rotating thin-walled isotropic circular cylindrical shell. Bleich & Baron [1954], Fung *et al.* [1957], Armenakas & Herrmann [1963], Mizoguchi [1963], Forsberg [1964] and Macke [1966] also made their contributions in this field.

Later works include the study of finite-length rotating cylindrical shell by Zohar & Aboudi [1973] who employed an exponential matrix expansion or fundamental matrix. Wang & Chen [1974] also presented a similar work on the free vibration of a rotating cylindrical shell. For the free vibration of a rotating cylindrical shell under initial stress, Penzes & Kraus [1972] used complex mode shapes to discuss the rotating cylindrical shells with general homogenous boundary conditions subject to initial normal pressure and axial load. Padovan [1973, 1975a,b] used, respectively, the complex Fourier series, numerical integration and finite element procedures to study similar problems. For the engineering application of electromagnetic shields in electrical machines for power generation, Shevchuk & Thullen [1978] produced experimentally the traveling-wave phenomena in a rotating cylindrical shell by means of a stationary exciting force. For the forced vibration of a rotating cylindrical shell, Fox & Hardie [1985] studied a harmonic and radial excitation acting at a fixed point on the shell based on the Flügge thin-shell theory. Huang & Soedel [1987a,b, 1988a,b] also discussed other different types of excitations including harmonic, periodic or distributed loads using a modal trigonometric expansion technique. Saito *et al.* [Saito & Endo, 1986a–c; Saito *et al.,* 1986, 1989] used the Flügge thin-shell theory to analyze the influence of boundary conditions in rotating shells. In addition, Haughton [1982] employed the membrane theory to investigate the small amplitude vibration of finitely deformed rotating elastic membrane cylinders composed of compressible isotropic hyperelastic material.

Recent works by Huang & Hsu [1990, 1992] include the studies of a rotating ring-stiffened thin cylindrical shell with multi-ring stiffeners and the resonance of a rotating cylindrical shell due to the action of harmonic moving loads. For rotating shells with variable thickness, Suzuki *et al.* [1991, 1993] used power series expansions to analyze the free vibration of a rotating circular cylindrical shell with axial quadratic thickness variation. Simha *et al.* [1994] studied a rotating shallow shell with variable thickness by treating the product of density and thickness as a single variable and then using FEM. For a moderately thick rotating circular cylindrical shell, Sivadas & Ganesan [1994] analyzed the natural frequency and damping factor using FEM and moderately thick-shell theory with shear deformation and rotatory inertia. Moreover, the increasing applications of composite structures have led to further investigations on the rotating composite cylindrical shell. They include Rand & Stavsky [1991] who used Love's shell theory and Wang & Lin [1993] who made a stress analysis including interfacial stresses. Chun & Bert [1993] also investigated a rotating circular cylindrical hollow shaft with layers of arbitrarily laminated composite materials using both thin-shell and thick-shell theories. For the consideration of large deformation and Coriolis acceleration, Chen *et al.* [1993] established the general vibrational governing equations for the rotating shell of revolution by linear approximation and analyzed the problem using FEM. Studies have been presented on the rotating composite and sandwich-type cylindrical shells [Lam & Loy, 1994, 1995a, 1998; Loy *et al.*, 1999] and on a comparison of different thin-shell theories for a rotating cylindrical shell [Lam & Loy, 1995b]. Influence of boundary conditions on the frequency characteristics of the rotating cylindrical shell by the GDQ method has also been discussed by Li & Lam [1998]. Other notable works include those carried out by Zinberg & Symonds [1970], Maewal [1981], dos Reis *et al.* [1987], Koga [1988], Khdeir *et al.* [1989], Smirnov [1989], Kobayashi & Yamada [1991], Nosier & Reddy [1992], Koff & El-Aini [1993], Sivadas & Ganesan [1993], Cai [1994], Ng *et al.* [1998], Ng & Lam [1999] and Ng [2003].

In this chapter, the important aspects of the study on the free vibration of thin rotating cylindrical shells are summarized. Following the numerical validation of the vibrational frequencies computed by different thin-shell theories, detailed parametric studies are carried out to examine the influences of physical parameters on the frequency characteristics, namely, boundary condition, rotating velocity, geometric dimension and model wave number.

3.2   Theoretical development: rotating thin cylindrical shell.

Consider a thin circular cylindrical shell rotating about its symmetrical and horizontal axis at a constant angular speed Ω, as shown in Fig. 3.1. In the figure, R is the constant mean radius, while the length and the thickness of the cylindrical shell are denoted by L and h, respectively. The reference surface for the deformations of the rotating

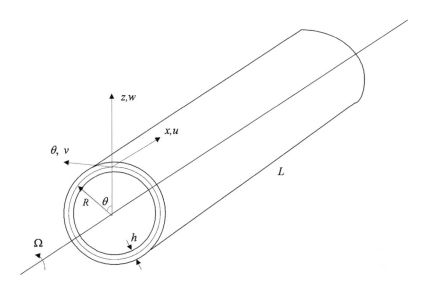

Figure 3.1 Geometry and coordinate system of a rotating cylindrical shell.

cylindrical shell is taken to be the middle surface on which an orthogonal coordinate system (x, θ, z) is fixed. Deformations of the rotating cylindrical shell are defined by u, v, and w in the longitudinal x, circumferential θ and normal z-directions, respectively.

From the definitions of the Lamé parameters, A_1 and A_2 can be dimensional or non-dimensional, depending on the dimensions of the curvilinear coordinates α_1 and α_2. For the present rotating cylindrical shell (the angle in the α_1-direction $\varphi = 90°$), if the transformations of the curvilinear coordinate systems are taken as

$$\alpha_1 = x \qquad \text{and} \qquad \alpha_2 = \theta \tag{3.1}$$

the Lamé parameters are

$$A_1 = 1 \qquad \text{and} \qquad A_2 = R \tag{3.2}$$

From the definition of the radii R_1 and R_2 described above, it is obvious that the radius R_1 is infinity, and the radius R_2 is a constant

$$R_1 = \infty \qquad \text{and} \qquad R_2 = R \tag{3.3}$$

Based on the above descriptions, substituting Eqs. (3.1)–(3.3) into Eqs. (2.62)–(2.67) and (2.68)–(2.73), two sets of resulting equations are obtained, respectively, for the basic and additional states of the present rotating cylindrical shell.

For the basic state of the rotating cylindrical shell, the resulting set of equations can be simplified and expressed as

$$R\frac{\partial N_{Bx}}{\partial x} + \frac{\partial N_{B\theta x}}{\partial \theta} = 0 \tag{3.4}$$

$$R\frac{\partial N_{Bx\theta}}{\partial x} + \frac{\partial N_{B\theta}}{\partial \theta} + Q_{B\theta} = 0 \tag{3.5}$$

$$R\frac{\partial Q_{Bx}}{\partial x} + \frac{\partial Q_{B\theta}}{\partial \theta} - N_{B\theta} + \rho h \Omega^2 R^2 = 0 \tag{3.6}$$

$$R\frac{\partial M_{Bx}}{\partial x} + \frac{\partial M_{B\theta x}}{\partial \theta} - Q_{Bx}R = 0 \tag{3.7}$$

$$R\frac{\partial M_{Bx\theta}}{\partial x} + \frac{\partial M_{B\theta}}{\partial \theta} - Q_{B\theta}R = 0 \tag{3.8}$$

$$M_{B\theta x} - r(x)(N_{Bx\theta} - N_{B\theta x}) = 0 \tag{3.9}$$

It is observed that no time component is included in the above Eqs. (3.4)–(3.9). As a result, the basic state equations of the rotating cylindrical shell can be taken as the initial state equations for the free vibration of the shell. For ease of formulation, free boundary conditions are imposed on both edges of the shell, $x = 0$ and L, for the basic state equations. In this way, the solutions of the initial state equations, i.e., those of the basic state equations, are obtained below for free vibration of the rotating shell

$$N_{B\theta} = \rho h \Omega^2 R^2, \qquad N_{Bx} = N_{Bx\theta} = N_{B\theta x} = 0 \tag{3.10}$$

$$M_{Bx} = M_{B\theta} = M_{Bx\theta} = M_{B\theta x} = 0, \qquad Q_{Bx} = Q_{B\theta} = 0 \tag{3.11}$$

The only nonzero solution $N_{B\theta}$ denotes the hoop tension due to the rotating velocity Ω. It is expressed by N_θ^0 and is defined as the initial hoop tension due to the centrifugal force effect

$$N_\theta^0 = \rho h \Omega^2 R^2 \tag{3.12}$$

Furthermore, for the additional state of the rotating cylindrical shell, using the solutions of the basic state Eqs. (3.10)–(3.12), the set of resulting equations is written as

$$\frac{1}{R}\left(R\frac{\partial N_{Ax}}{\partial x} + \frac{\partial N_{A\theta x}}{\partial \theta}\right) + \rho h \Omega^2 R\vartheta - \rho h \frac{\partial^2 u}{\partial t^2} = 0 \tag{3.13}$$

$$\frac{1}{R}\left(R\frac{\partial N_{Ax\theta}}{\partial x} + \frac{\partial N_{A\theta}}{\partial \theta} + Q_{A\theta}\right) + \frac{N_\theta^0}{R}\left(\frac{\partial e_1}{\partial \theta} - \psi\right) + \rho h \Omega^2 R\psi - 2\rho h \Omega \frac{\partial w}{\partial t}$$

$$- \rho h \frac{\partial^2 v}{\partial t^2} = 0 \tag{3.14}$$

$$\frac{1}{R}\left(R\frac{\partial Q_{Ax}}{\partial x} + \frac{\partial Q_{A\theta}}{\partial \theta} - \frac{\partial(N_\theta^0 \psi)}{\partial \theta} - N_{A\theta} + N_\theta^0 e_2\right) - \rho h \Omega^2 R e_1 + 2\rho h \Omega \frac{\partial w}{\partial t}$$

$$- \rho h \frac{\partial^2 w}{\partial t^2} = 0 \tag{3.15}$$

$$\frac{1}{R}\left(R\frac{\partial M_{Ax}}{\partial x} + \frac{\partial M_{A\theta x}}{\partial \theta}\right) - Q_{Ax} = 0 \tag{3.16}$$

$$\frac{1}{R}\left(R\frac{\partial M_{Ax\theta}}{\partial x} + \frac{\partial M_{A\theta}}{\partial \theta}\right) - Q_{A\theta} = 0 \tag{3.17}$$

$$\frac{M_{A\theta x}}{R} - N_{Ax\theta} + N_{A\theta x} = 0 \tag{3.18}$$

From Eqs. (3.16) and (3.17) we have

$$Q_{Ax} = \frac{1}{R}\left(R\frac{\partial M_{Ax}}{\partial x} + \frac{\partial M_{A\theta x}}{\partial \theta}\right) \tag{3.19}$$

$$Q_{A\theta} = \frac{1}{R}\left(R\frac{\partial M_{Ax\theta}}{\partial x} + \frac{\partial M_{A\theta}}{\partial \theta}\right) \tag{3.20}$$

From the general strain–displacement relationships of shells of revolution defined earlier in Eqs. (2.46)–(2.48), and considering Eqs. (3.1)–(3.3), the Love type of geometric deformation relationships for the rotating cylindrical shell can be obtained as

$$e_1 = \frac{\partial u}{\partial x}, \qquad e_2 = \frac{1}{R}\frac{\partial v}{\partial \theta} + \frac{w}{R}, \qquad \vartheta = -\frac{\partial w}{\partial x}, \qquad \psi = -\frac{1}{R}\frac{\partial w}{\partial \theta} + \frac{v}{R} \tag{3.21}$$

$$e_{12} = \beta_1 + \beta_2, \qquad \beta_1 = \frac{\partial v}{\partial x}, \qquad \beta_2 = \frac{1}{R}\frac{\partial u}{\partial \theta}, \tag{3.22}$$

$$\kappa_x = -\frac{\partial^2 w}{\partial x^2}, \qquad \kappa_\theta = \frac{1}{R^2}\left(-\frac{\partial^2 w}{\partial \theta^2} + \frac{\partial v}{\partial \theta}\right), \qquad \kappa_{x\theta} = \frac{2}{R}\left(-\frac{\partial^2 w}{\partial x\,\partial\theta} + \frac{\partial v}{\partial x}\right) \tag{3.23}$$

By substituting the resulting Eqs. (3.19)–(3.23) into Eqs. (3.13)–(3.15) and simplifying, the Love-type governing equations of motion in terms of the force and moment resultants and displacements are finally derived, for a rotating cylindrical shell and they are expressed as follows,

$$\frac{\partial N_x}{\partial x} + \frac{1}{R}\frac{\partial N_{x\theta}}{\partial \theta} + N_\theta^0\left(\frac{1}{R^2}\frac{\partial^2 u}{\partial \theta^2} - \frac{1}{R}\frac{\partial w}{\partial x}\right) - \rho h \frac{\partial^2 u}{\partial t^2} = 0 \tag{3.24}$$

$$\frac{\partial N_{x\theta}}{\partial x} + \frac{1}{R}\frac{\partial N_\theta}{\partial \theta} + \frac{1}{R}\frac{\partial M_{x\theta}}{\partial x} + \frac{1}{R^2}\frac{\partial M_\theta}{\partial \theta} + \frac{N_\theta^0}{R}\frac{\partial^2 u}{\partial x\,\partial\theta}$$

$$- \rho h\left(\frac{\partial^2 v}{\partial t^2} + 2\Omega\frac{\partial w}{\partial t} - \Omega^2 v\right) = 0 \tag{3.25}$$

$$\frac{\partial^2 M_x}{\partial x^2} + \frac{2}{R}\frac{\partial^2 M_{x\theta}}{\partial x\,\partial\theta} + \frac{1}{R^2}\frac{\partial^2 M_\theta}{\partial \theta^2} - \frac{N_\theta}{R} + \frac{N_\theta^0}{R^2}\left(\frac{\partial^2 w}{\partial \theta^2} - \frac{\partial v}{\partial \theta}\right)$$

$$- \rho h\left(\frac{\partial^2 w}{\partial t^2} - 2\Omega\frac{\partial v}{\partial t} - \Omega^2 w\right) = 0 \tag{3.26}$$

where $N_\theta^0 = \rho h \Omega^2 R^2$ is the initial hoop tension due to the centrifugal force effect. It is also noted that, in the above resulting expressions, the subscript A is omitted for ease of subsequent discussion.

Similarly, the different sets of the governing equations, corresponding to other thin-shell theories, can be derived for the rotating thin cylindrical shell. In order to unify the expressions of governing equations for various shell theories, a compact unified form of the governing equations of motion for Love's, Donnell's, Flügge's, and Sanders' shell theories may be written as follows:

$$\frac{\partial N_x}{\partial x} + \frac{1}{R}\frac{\partial N_{\theta x}}{\partial \theta} - \frac{\delta_1}{R}\frac{\partial M_{\theta x}}{\partial \theta} + N_\theta^0\left(\frac{1}{R^2}\frac{\partial^2 u}{\partial \theta^2} - \frac{1}{R}\frac{\partial w}{\partial x}\right) - \rho h \frac{\partial^2 u}{\partial t^2} = 0 \tag{3.27}$$

Table 3.1

Tracers appearing in Eqs. (3.27)–(3.29) for Donnell's, Flügge's, Love's and Sanders' shell theories.

Shell theory	Donnell	Flügge	Love	Sanders
δ_1	0	0	0	1/2
δ_2	0	1	1	3/2
δ_3	0	1	1	1

$$\frac{\partial N_{x\theta}}{\partial x} + \frac{1}{R}\frac{\partial N_\theta}{\partial \theta} + \frac{\delta_2}{R}\frac{\partial M_{x\theta}}{\partial x} + \frac{\delta_3}{R^2}\frac{\partial M_\theta}{\partial \theta} + \frac{N_\theta^0}{R}\frac{\partial^2 u}{\partial x\,\partial \theta}$$

$$- \rho h\left(\frac{\partial^2 v}{\partial t^2} + 2\Omega\frac{\partial w}{\partial t} - \Omega^2 v\right) = 0 \tag{3.28}$$

$$\frac{\partial^2 M_x}{\partial x^2} + \frac{1}{R}\left(\frac{\partial^2 M_{x\theta}}{\partial x\,\partial \theta} + \frac{\partial^2 M_{\theta x}}{\partial x\,\partial \theta}\right) + \frac{1}{R^2}\frac{\partial^2 M_\theta}{\partial \theta^2} - \frac{N_\theta}{R} + \frac{N_\theta^0}{R^2}\left(\frac{\partial^2 w}{\partial \theta^2} - \frac{\partial v}{\partial \theta}\right)$$

$$- \rho h\left(\frac{\partial^2 w}{\partial t^2} - 2\Omega\frac{\partial v}{\partial t} - \Omega^2 w\right) = 0 \tag{3.29}$$

where δ_1, δ_2 and δ_3 are tracers and their values are tabulated in Table 3.1.

3.3   Numerical implementation.

a)   Galerkin's method (characteristic beam functions).

In this section, the classical Galerkin's method is employed for numerical analysis of the frequency characteristics of thin rotating cylindrical shells. As a weighted-residual method based on the variational principle, Galerkin's method seeks the solution from a weighted-integral statement of the governing equations, and requires that the trial functions satisfy the given boundary conditions but not the differential governing equations of the problem. Unlike the Rayleigh–Ritz method, Galerkin's method does not require the formulation of an energy functional. Before using Galerkin's method, however, we need to obtain the governing equations in the form where the only variables are the displacements, and also construct the trial functions satisfying the given boundary conditions.

By selecting the geometric deformation relationships from Eqs. (2.2)–(2.13), corresponding to various shell theories, and substituting them into the constitutive relationship (2.34), followed by substituting the resulting equations into the governing equations (3.27)–(3.29) in terms of the forces, moments and displacements, a set of three-dimensional partial differential governing equations with constant coefficients in terms of the displacements u, v, and w is derived for the free vibration of a rotating cylindrical shell and can be written in the following form:

$$\bar{L}_{11}u + \bar{L}_{12}v + \bar{L}_{13}w = 0$$

$$\bar{L}_{21}u + \bar{L}_{22}v + \bar{L}_{23}w = 0 \tag{3.30}$$

$$\bar{L}_{31}u + \bar{L}_{32}v + \bar{L}_{33}w = 0$$

or in the form of vectors as

$$\bar{\mathbf{L}}\mathbf{U} = \mathbf{0} \tag{3.31}$$

where $\mathbf{U}^{\mathrm{T}} = \{u(x, \theta, t), v(x, \theta, t), w(x, \theta, t)\}$ is a vibrational displacement field vector of the rotating cylindrical shell. $\bar{\mathbf{L}} = [\bar{L}_{ij}]$ ($i, j = 1, 2, 3$) is a 3×3 differential operator matrix of $\mathbf{U}$. For example, corresponding to Love's shell theories, the differential operators $\bar{L}_{ij}$ ($i, j = 1, 2, 3$) are detailed in Appendix A.

For the analysis of the free vibration of thin rotating cylindrical shell, the assumed-mode method can be employed in a straightforward manner. Several common boundary conditions encountered in generic engineering analysis are considered here. They are the clamped boundary condition with zero transverse displacement and zero rotation about the edge, the simply supported boundary condition with zero transverse displacement and zero bending moment about the edge, and the free boundary condition without any constraint. As such, the trial functions for free vibration of a thin rotating cylindrical shell can be expressed in displacement-field form as follow:

$$u(x, \theta, t) = \sum_{m=1}^{\infty} \sum_{n=1}^{\infty} U_{mn}\phi_m^u(x)\varphi_n^u(\theta, t)$$

$$v(x, \theta, t) = \sum_{m=1}^{\infty} \sum_{n=1}^{\infty} V_{mn}\phi_m^v(x)\varphi_n^v(\theta, t) \tag{3.32}$$

$$w(x, \theta, t) = \sum_{m=1}^{\infty} \sum_{n=1}^{\infty} W_{mn}\phi_m^w(x)\varphi_n^w(\theta, t)$$

where U_{mn}, V_{mn} and W_{mn} are unknown displacement amplitudes, m and n are integers representing the axial (longitudinal) and circumferential wave numbers of vibrational cylindrical shell, respectively. For the present thin circular rotating cylindrical shell, the functions $\phi_m^k(x)$ ($k = u$, v, w) are the axial modal functions describing the vibrational mode in longitudinal direction, which are required to satisfy at least the geometric boundary conditions. $\varphi_n^k(\theta, t)$ ($k = u$, v, w) are the circumferential modal functions describing the circumferential modes and can be taken in the following form,

$$\varphi_n^u(\theta, t) = \varphi_n^w(\theta, t) = \cos(n\theta + \omega t) \qquad \text{and} \qquad \varphi_n^v(\theta, t) = \sin(n\theta + \omega t) \qquad (3.33)$$

where ω (rad/s) is the natural circular frequency of the rotating cylindrical shell.

Moreover, in the axial direction of the rotating cylindrical shell, the axial modal functions $\phi_m^k(x)$ ($k = u$, v, w) are written as the characteristic beam functions which can be expressed in a general form as

$$\phi_m^k(x) = \left[\zeta_1 \sin\left(\frac{\lambda_m x}{L}\right) + \zeta_2 \sinh\left(\frac{\lambda_m x}{L}\right)\right] + \zeta_m\left[\zeta_3 \cos\left(\frac{\lambda_m x}{L}\right) + \zeta_4 \cosh\left(\frac{\lambda_m x}{L}\right)\right] \qquad (3.34)$$

It is noted that the boundary conditions of common beams with isotropic material and constant area moment of inertia for the clamped, simply supported, and free-type of boundary condition can be expressed mathematically as

clamped boundary:

$$\varphi_m(x) = \frac{\partial \varphi_m(x)}{\partial x} = 0 \qquad (3.35)$$

simply supported boundary:

$$\varphi_m(x) = \frac{\partial^2 \varphi_m(x)}{\partial x^2} = 0 \qquad (3.36)$$

free boundary:

$$\frac{\partial^2 \varphi_m(x)}{\partial x^2} = \frac{\partial^3 \varphi_m(x)}{\partial x^3} = 0 \qquad (3.37)$$

Thus, the coefficients ζ_1, ζ_2, ζ_3, ζ_4 can be written and ζ_m is computed as the roots of some transcendental equations relevant to λ_m for the various boundary condition combinations as follows:

simply supported–simply supported

$$\lambda_m = m\pi, \qquad \zeta_1 = 1, \qquad \zeta_2 = \zeta_3 = \zeta_4 = \zeta_m = 0 \tag{3.38}$$

free–free

$$\cos \lambda_m \cosh \lambda_m = 1, \qquad \zeta_1 = \zeta_2 = \zeta_3 = \zeta_4 = 1, \qquad \zeta_m = \frac{\sin \lambda_m - \sinh \lambda_m}{\cosh \lambda_m - \cos \lambda_m} \tag{3.39}$$

clamped–clamped

$$\cos \lambda_m \cosh \lambda_m = 1, \qquad \zeta_1 = \zeta_3 = -1, \qquad \zeta_2 = \zeta_4 = 1,$$

$$\zeta_m = \frac{\sinh \lambda_m - \sin \lambda_m}{\cos \lambda_m - \cosh \lambda_m} \tag{3.40}$$

clamped–free

$$\cos \lambda_m \cosh \lambda_m = -1, \qquad \zeta_1 = \zeta_3 = 1, \qquad \zeta_2 = \zeta_4 = -1,$$

$$\zeta_m = -\left(\frac{\sin \lambda_m + \sinh \lambda_m}{\cos \lambda_m + \cosh \lambda_m} \right) \tag{3.41}$$

clamped–simply supported

$$\tan \lambda_m = \tanh \lambda_m, \qquad \zeta_1 = \zeta_4 = 1, \qquad \zeta_2 = \zeta_3 = -1,$$

$$\zeta_m = \frac{\sin \lambda_m - \sinh \lambda_m}{\cos \lambda_m - \cosh \lambda_m} \tag{3.42}$$

For the thin circular rotating cylindrical shell, based on the characteristic beam functions expressed by Eq. (3.34), the trial functions for the free vibration of the shell can be constructed in displacement-field form expressed by Eq. (3.32) through the combination of the various boundary conditions of the characteristic beam functions mentioned above. They are able to satisfy the given boundary conditions, or at least the geometric boundary conditions, for the rotating cylindrical shell.

For the partial differential governing equations (3.30) of motion of the thin rotating cylindrical shell, the weighted-integral statement of Galerkin's method can be written mathematically as follows:

$$\int_t \int_\theta \int_x (\bar{L}_{11}u + \bar{L}_{12}v + \bar{L}_{13}w)u \ dx \ d\theta \ dt = 0$$

$$\int_t \int_\theta \int_x (\bar{L}_{21}u + \bar{L}_{22}v + \bar{L}_{23}w)v \ dx \ d\theta \ dt = 0 \qquad (3.43)$$

$$\int_t \int_\theta \int_x (\bar{L}_{31}u + \bar{L}_{32}v + \bar{L}_{33}w)w \ dx \ d\theta \ dt = 0$$

where u, v, and w are the trial functions of displacement field expressed by Eq. (3.32) for the free vibration of the rotating cylindrical shell. They include the undetermined parameters that can be obtained from the integral computation of the simultaneous equations (3.43).

Substituting the trial functions of displacement fields (3.32) into the weighted-integral statement (3.43) of Galerkin's method, and then performing integration in the resulting expressions, the eigenvalue equation for the free vibration of rotating cylindrical shell can be obtained and written in the following matrix form:

$$\begin{bmatrix} C_{11} & C_{12} & C_{13} \\ C_{21} & C_{22} & C_{23} \\ C_{31} & C_{32} & C_{33} \end{bmatrix} \begin{Bmatrix} U_{mn} \\ V_{mn} \\ W_{mn} \end{Bmatrix} = \begin{Bmatrix} 0 \\ 0 \\ 0 \end{Bmatrix} \qquad (3.44)$$

where C_{ij} $(i, j = 1, 2, 3)$ are the coefficients in terms of natural circular frequency, material elastic constants and geometric parameters and so on. It is noted that the above eigenvalue equation (3.44) for the free vibration of rotating cylindrical shell can be transformed mathematically and rewritten into the following form:

$$\left(\omega^2 \begin{bmatrix} C_{11}^* & 0 & 0 \\ 0 & C_{22}^* & 0 \\ 0 & 0 & C_{33}^* \end{bmatrix} + \omega \begin{bmatrix} C_{11}^{**} & C_{12}^{**} & C_{13}^{**} \\ C_{21}^{**} & C_{22}^{**} & C_{23}^{**} \\ C_{31}^{**} & C_{32}^{**} & C_{33}^{**} \end{bmatrix} + \begin{bmatrix} C_{11}^{***} & C_{12}^{***} & C_{13}^{***} \\ C_{21}^{***} & C_{22}^{***} & C_{23}^{***} \\ C_{31}^{***} & C_{32}^{***} & C_{33}^{***} \end{bmatrix} \right)$$
$$\begin{Bmatrix} U_{mn} \\ V_{mn} \\ W_{mn} \end{Bmatrix} = \begin{Bmatrix} 0 \\ 0 \\ 0 \end{Bmatrix} \qquad (3.45)$$

where coefficients C_{ii}^*, C_{ij}^{**} and C_{ij}^{***} $(i, j = 1, 2, 3)$ are computed from the re-arrangement of C_{ij} $(i, j = 1, 2, 3)$ in Eq. (3.44).

If the zero rotating velocity $\Omega = 0$ is taken in formulation, corresponding eigenvalue equation for the free vibration of a corresponding stationary cylindrical shell can be written as

$$\left(\omega^2 \begin{bmatrix} C_{11}^* & 0 & 0 \\ 0 & C_{22}^* & 0 \\ 0 & 0 & C_{33}^* \end{bmatrix} + \begin{bmatrix} C_{11}^{***} & C_{12}^{***} & C_{13}^{***} \\ C_{21}^{***} & C_{22}^{***} & C_{23}^{***} \\ C_{31}^{***} & C_{32}^{***} & C_{33}^{***} \end{bmatrix} \right) \begin{Bmatrix} U_{mn} \\ V_{mn} \\ W_{mn} \end{Bmatrix} = \begin{Bmatrix} 0 \\ 0 \\ 0 \end{Bmatrix} \tag{3.46}$$

Comparing the eigenvalue equations (3.45) and (3.46), it can be seen that, because Eq. (3.46) is a standard eigenvalue equation, the eigensolution of free vibration for non-rotating shell can be easily obtained using a general eigenvalue approach. However, for rotating shells, Eq. (3.45) is a non-standard eigenvalue equation such that the eigensolution of free vibration cannot be obtained directly unless mathematical transformations are performed.

By imposing the non-trivial solution condition, and then setting the determinant of the characteristic matrix equal to zero, the eigenvalue equation (3.44) can be solved as

$$\begin{vmatrix} C_{11} & C_{12} & C_{13} \\ C_{21} & C_{22} & C_{23} \\ C_{31} & C_{32} & C_{33} \end{vmatrix} = 0 \tag{3.47}$$

Expanding the above determinant (3.47), a polynomial equation with respect to the natural circular frequency ω is obtained

$$d_0\omega^6 + d_1\omega^4 + d_2\omega^3 + d_3\omega^2 + d_4\omega + d_5 = 0 \tag{3.48}$$

where d_i $(i = 0, 1, 2...6)$ are constant coefficients. The six roots in Eq. (3.48) can be found using standard Newton–Raphson procedure. For the special case of a very long cylindrical shell $(L \to \infty)$, the smallest coefficient among d_3, d_4 and d_5 is larger than d_0, d_1 and d_2 by an order of 10^5. Thus the coefficients d_0, d_1 and d_2 can be neglected for this special case. Hence, for a very long rotating cylindrical shell, the frequency equation is reduced to

$$d_3\omega^2 + d_4\omega + d_5 = 0 \tag{3.49}$$

In the solutions of the eigenvalue equations mentioned above, there are the two roots whose absolute values are the smallest real numbers, one positive and the other negative.

Both the eigenvalues are the eigensolution of frequency characteristics of the rotating cylindrical shell, corresponding, respectively, to the backward and forward traveling waves or to the positive and negative rotating velocities. The positive eigenvalue correspond to the backward wave due to a rotation $\Omega > 0$; and the negative one correspond to the forward wave due to a rotation $\Omega < 0$. Thus the frequency solution of the rotating cylindrical shell is related to the rotating velocity for each vibration mode expressed by a pair of wave numbers (m, n), where m and n are the longitudinal and circumferential wave numbers, respectively. If the cylindrical shell is stationary ($\Omega = 0$), the magnitudes of both the eigenvalues are identical, corresponding to a standing wave vibration. Once the shell starts to rotate, however, the standing wave motion will be transformed to backward or forward waves, depending on the direction of rotation. It is generally observed from the computed frequencies that the absolute values of the backward waves are larger than those of forward waves.

b)   *Convergence characteristics and numerical validation.*

In order to examine the computational accuracy of Galerkin's method whose trial functions are constructed by the characteristic beam functions, several numerical comparisons of the presently computed frequency characteristics are made with those reported in published literature. The comparisons in this section are, respectively, for stationary cylindrical and conical shells with various boundary conditions and an infinitely long rotating cylindrical shell. The comparison for a finite-length rotating cylindrical shell is not presented because numerical results for this subject are not available in the literature, where the prior relevant work presented only limited graphical results with no documentation of any numerical data.

To investigate the vibration of stationary isotropic cylindrical shells with different boundary conditions, the numerical comparisons are presented in Table 3.2 for the case of simply supported boundary conditions at both edges, in Tables 3.3 and 3.4 for fully clamped boundary conditions, and in Table 3.5 for clamped-free boundary conditions. In Table 3.2 the comparison is against the results of Markus [1988], where the present Galerkin results are compared with solutions obtained by three-dimensional elasticity theory and Flügge's theory, respectively. Table 3.3 shows the comparisons of the present Galerkin results with the exact solutions of Dym [1973] and the series solutions of Chung [1981]. In Tables 3.2 and 3.3, the non-dimensional frequency parameter $f^* = \omega R \sqrt{(1 - \mu^2)\rho/E}$ is used in the comparisons, in which ω (rad/s) is the natural circular frequency, R is the radius, μ is Poisson's ratio, ρ is the density and E is Young's modulus of elasticity. In Tables 3.4 and 3.5, however, the natural frequency (Hz) computed is compared directly with the experimental results of Koval & Cranch [1962] and Sharma

Table 3.2

Comparison of frequency parameter $f^* = \omega R \sqrt{(1 - \mu^2)\rho/E}$ for non-rotating isotropic cylindrical shell with simply supported boundary conditions at both edges ($\mu = 0.3, m = 1$).

			Markus [1988]		Galerkin's method
R/L	h/R	n	Flügge's theory	3D solutions	(Love's theory)
0.05	0.05	0	0.0961909	0.0929296	0.0929682
		1	0.0163488	0.0161063	0.0161029
		2	0.0392916	0.0392332	0.0392710
		3	0.109782	0.109477	0.1098113
		4	0.210197	0.209008	0.2102770
0.05	0.002	0	0.096191	0.0929296	0.0929298
		1	0.0163518	0.0161011	0.0161011
		2	0.00552868	0.00545243	0.00545297
		3	0.00506018	0.00503724	0.00504148
		4	0.00853766	0.00853409	0.00853383
4	0.002	0	0.962321	0.957994	0.958014
		1	0.956227	0.951993	0.952010
		2	0.938425	0.934462	0.934474
		3	0.910267	0.906734	0.906746
		4	0.873735	0.870765	0.870776

[1973], respectively. The above comparisons demonstrate that the present frequency results simulated by Galerkin's method agree well with those in the open literature except in Table 3.4 for $n = 5$, a difference of 9.1% between the computational and experimental results is observed. This is because of the difficulty in precisely simulating the very rigid clamped–clamped boundary conditions in the experiment, where the influence is significant for small circumferential wave number.

Table 3.3

Comparison of frequency parameter $f^* = \omega R \sqrt{(1 - \mu^2)\rho/E}$ for non-rotating single-layer isotropic cylindrical shell with clamped–clamped boundary conditions ($\mu = 0.3, m = 1$).

Case	Dym [1973] (exact solution)	Chung [1981] (series solution)	Present Galerkin's method (Love's theory)
$L/R = 10, R/h = 500, n = 4$	0.01508	0.01515	0.01512
$L/R = 10, R/h = 20, n = 2$	0.05784	0.05795	0.05789
$L/R = 2, R/h = 20, n = 3$	0.3118	0.3119	0.3119

Table 3.4

Comparison of natural frequency (Hz) for a non-rotating isotropic
circular cylindrical shell with clamped–clamped boundary conditions
($m = 1$, $L = 12$ in., $R = 3$ in., $h = 0.01$ in., $\mu = 0.3$,
$E = 30 \times 10^6$ lb/in.^{2}).

n	Koval & Cranch [1962]	Galerkin's method (Love's theory)	Relative error (%)
5	552	602.2	9.1
6	525	547.4	4.2
7	592	601	1.4
8	720	723.8	0.4
9	885	890	0.6
10	1095	1087.4	0.7

For the stationary orthotropic cylindrical shells, the validation is carried out for a cylindrical shell with the simply supported boundary conditions at both edges. Comparisons for other boundary conditions are not presented as the results in form of numerical data for other boundary conditions are not available in the published literature. The comparison is presented in Table 3.6 for a two-layered cross-ply [90°/0°] circular cylindrical shell. The material properties used are $E_{11}/E_{22} = 40$, $G_{12}/E_{22} = 0.5$ and $\mu_{12} = 0.25$. The present frequency parameter $f^* = \omega L_s\sqrt{\rho h/A_{11}}$ computed by Galerkin's method is compared with the solution of Soldatos [1984], in which ω is the natural circular frequency, $L_s = 2\pi R$ where R is the radius, ρ is the density, h is the thickness and A_{11} is an extensional stiffness term defined in Eq. (2.35). The comparison shows that the present frequency results are in good agreement with those in the literature.

Table 3.5

Comparison of natural frequency (Hz) for a non-rotating isotropic circular cylindrical
shell with clamped–free boundary conditions ($m = 1$, $L = 502$ mm, $R = 63.5$ mm,
$h = 1.63$ mm, $\mu = 0.28$, $E = 2.1 \times 10^{11}$ N/m^2, $\rho = 7.8 \times 10^3$ kg/m^3).

n	Sharma [1973]	Galerkin's method (Love's theory)	Relative error (%)
3	760.0	769.9	1.3
4	1451.0	1459.7	0.6
5	2336.0	2366.9	1.3
6	3429.0	3470.0	1.8

Table 3.6

Comparison of frequency parameter $f^* = \omega L_s \sqrt{\rho h / A_{11}}$ for a non-rotating two-layered cross-ply [90°/0°] circular cylindrical shell with simply supported boundary conditions at both edges ($h/R = 0.01$).

L_s/R	n	Soldatos [1984]	Galerkin's method (Love's theory)
1.0	1	2.106	2.106
	2	1.344	1.344
	3	0.9589	0.9586
	4	0.7495	0.7492
	5	0.6423	0.6419
	6	0.6138	0.6134
2.0	1	1.073	1.073
	2	0.6710	0.6709
	3	0.4710	0.4709
	4	0.3774	0.3772
	5	0.3632	0.3631
	6	0.4168	0.4166

For the non-rotating isotropic conical shells with different boundary conditions, two comparisons are made, in Table 3.7 for the case of simply supported boundary condition at both edges and in Table 3.8 for the fully clamped boundary condition case. The present Galerkin results based on characteristic beam functions are compared against the solution

Table 3.7

Comparison of frequency parameter $f = \omega b \sqrt{((1 - \mu^2)\rho/E)}$ for non-rotating conical shell with simply supported boundary conditions at both edges ($m = 1$, $\mu = 0.3$, $h/b = 0.01$, $L \sin \alpha / b = 0.25$).

n	$\alpha = 30°$ Present[a]	$\alpha = 30°$ Irie *et al.* [1984]	$\alpha = 45°$ Present[a]	$\alpha = 45°$ Irie *et al.* [1984]	$\alpha = 60°$ Present[a]	$\alpha = 60°$ Irie *et al.* [1984]
2	0.8420	0.7910	0.7655	0.6879	0.6348	0.5722
3	0.7376	0.7284	0.7212	0.6973	0.6238	0.6001
4	0.6362	0.6352	0.6739	0.6664	0.6145	0.6054
5	0.5528	0.5531	0.6323	0.6304	0.6111	0.6077
6	0.4950	0.4949	0.6035	0.6032	0.6171	0.6159
7	0.4661	0.4653	0.5921	0.5918	0.6350	0.6343
8	0.4660	0.4654	0.6001	0.5992	0.6660	0.6650
9	0.4916	0.4892	0.6273	0.6257	0.7101	0.7084

Results of Irie *et al.* [1984] were obtained by the numerical integration method

[a] Present results were obtained by Galerkin's method and based on characteristic beam functions

Table 3.8

Comparison of frequency parameter $f = \omega b \sqrt{((1 - \mu^2)\rho/E)}$ for non-rotating conical shell with clamped boundary conditions at both edges ($m = 1$, $\mu = 0.3$, $h/b = 0.01$, $L \sin \alpha/b = 0.5$).

	$\alpha = 45°$		$\alpha = 60°$	
n	Present[a]	Irie et al. [1984]	Present[a]	Irie et al. [1984]
1	0.8452	0.8120	0.6449	0.6316
2	0.6803	0.6696	0.5568	0.5523
3	0.5553	0.5430	0.4818	0.4785
4	0.4778	0.4570	0.4361	0.4298
5	0.4395	0.4095	0.4202	0.4093

Results of Irie et al. [1984] were obtained by using the numerical integration method
[a] Present results were obtained by Galerkin's method and based on characteristic beam functions

of Irie et al. [1984], which used the numerical integration method. In both these tables, the non-dimensional frequency parameter $f = \omega b \sqrt{((1 - \mu^2)\rho/E)}$ is used in the comparisons, in which ω is the natural circular frequency, b is the radius at the large edge of conical shell, μ is Poisson's ratio, ρ is the density and E is Young's modulus of elasticity. It is observed from both tables that, for different cone angles α, the correspondence of the two sets of frequency parameters are quite acceptable.

In the final comparison of this section, the earlier presented methodology in this chapter is validated for a rotating cylindrical shell. The comparison is conducted for an infinitely long isotropic rotating cylindrical shell against the results obtained by Eq. (45) of Chen et al. [1993]. This comparison is shown in Table 3.9 for two rotational velocities $\Omega = 0.05$ and 0.1 rps (rps, revolutions per second or Hz), and for parameters $h/R = 0.002$ and Poisson's ratio $\mu = 0.3$. The present results are obtained by reducing the unified analysis to Love's shell theory. The non-dimensional frequency parameters, $f_b^*, f_f^* = \omega R \sqrt{(1 - \mu^2)\rho/E}$, used in the comparisons represent the frequency parameters of the backward and forward waves, respectively. The comparison demonstrates that the present Galerkin results agree well with those in the literature for an infinitely long rotating isotropic cylindrical shell. Furthermore, for a very long rotating cylindrical shell, the frequency equation (3.48) can be simplified and reduced to the simpler frequency equation (3.49). For a numerical validation of this simplification, Table 3.10 is generated for an infinitely long rotating laminated orthotropic composite cylindrical shell. This rotating cylindrical shell is a three equithickness layered orthotropic shell with a stacking sequence $[90°/0°/90°]$ having a thickness ratio of $h/R = 0.002$. The relevant material properties of each layer are $E_{22} = 7.6$ GN/m^2, $E_{11}/E_{22} = 2.5$, $G_{12} = 4.1$ GN/m^2, $\mu_{12} = 0.26$, $\rho = 1643$ kg/m^3. It is observed from the table that the frequency results obtained by the frequency equation (3.48) and the simplified frequency equation (3.49) are in very good agreement.

Table 3.9

Comparison of frequency parameters $f_b^*, f_f^* = \omega R\sqrt{(1 - \mu^2)\rho/E}$ for an infinitely long rotating cylindrical shell ($m = 1$, $\mu = 0.3$, $h/R = 0.002$).

		Chen et al. [1993][a]		Present Galerkin's method	
Ω (rps)	n	f_b^*	f_f^*	f_b^*	f_f^*
0.05	2	0.00167	0.00142	0.00170	0.00145
	3	0.00448	0.00429	0.00450	0.00431
	4	0.00848	0.00833	0.00850	0.00835
	5	0.01370	0.01353	0.01367	0.01355
0.1	2	0.00180	0.00130	0.00189	0.00139
	3	0.00457	0.00419	0.00465	0.00428
	4	0.00855	0.00826	0.00863	0.00834
	5	0.01371	0.01347	0.01379	0.01355

The subscripts "b" and "f" denote the backward and forward waves, respectively
[a] From Eq. (45) of Chen et al. [1993]:

$$\omega_b^* = \frac{2n}{n^2 + 1}\Omega + \sqrt{\frac{n^2(n^2 - 1)^2}{n^2 + 1}\frac{Eh^2}{12(1 - \mu^2)\rho R^2} + \frac{n^4 + 3}{(n^2 + 1)^2}\Omega^2}$$

$$\omega_f^* = \frac{2n}{n^2 + 1}\Omega - \sqrt{\frac{n^2(n^2 - 1)^2}{n^2 + 1}\frac{Eh^2}{12(1 - \mu^2)\rho R^2} + \frac{n^4 + 3}{(n^2 + 1)^2}\Omega^2}$$

Table 3.10

Comparison of natural frequencies $\bar{f}^*$ (Hz) between the frequency equation (3.48) and simplified frequency equation (3.49) for an infinitely long rotating laminated composite cylindrical shell ($m = 1$, $h/R = 0.002$).

		Eq. (3.48)		Eq. (3.49)	
Ω (rps)	n	$\bar{f}_b^*$	$\bar{f}_f^*$	$\bar{f}_b^*$	$\bar{f}_f^*$
0.05	2	1.505924	0.705922	1.505918	0.705921
	3	3.010081	2.410077	3.010053	2.410061
	4	5.165541	4.694948	5.165459	4.694882
	5	7.964535	7.579912	7.964338	7.579738
1.0	2	2.459237	0.859233	2.459213	0.859231
	3	4.102077	2.902067	4.102010	2.902038
	4	6.328205	5.387015	6.328056	5.386914
	5	9.163919	8.394671	9.163623	8.394431

The subscripts "b" and "f" denote the backward and forward waves, respectively

3.4   Frequency characteristics.

a)   Influence of Coriolis and centrifugal effects.

As mentioned earlier in the section, "Galerkin's method (characteristic beam functions)", the standing wave motion is transformed when the cylindrical shell starts to rotate. Depending on the rotating direction, backward or forward traveling waves will emerge. Two rotational effects are present in a rotating shell, one is the Coriolis effect and the other is the centrifugal effect. These effects are found to have different influences on the natural frequencies of the rotating shell, which will be discussed in this section.

To examine the different influences of the Coriolis and centrifugal effects on the frequency characteristics of the rotating cylindrical shell, Fig. 3.2 is plotted for a multi-layered cylindrical shell with the simply supported boundary conditions at both edges, and which has three isotropic layers of construction. The thickness of the middle layer ($E = 2.0685 \times 10^{11}$ N/m^2, $\mu = 0.3$ and $\rho = 8053$ kg/m^3) is three times as thick as the other two surface layers ($E = 4.8265 \times 10^9$ N/m^2, $\mu = 0.3$ and $\rho = 1314$ kg/m^3). It shows the variation of the normalized natural frequencies f/f_0 with the rotating speed for two cases. In the first case, both the Coriolis and centrifugal effects are considered, and in the second case, only the centrifugal effect is considered. In the plot, f represents the natural frequencies of the backward and forward waves for the rotating cylindrical shell, and it is normalized with respect to the corresponding natural frequencies f_0 of the standing wave for a corresponding non-rotating cylindrical shell. The figure illustrates the different influences of the Coriolis and centrifugal effects on the frequency characteristics of a rotating shell. It is observed that, if both the Coriolis and centrifugal effects are considered, the natural frequencies of backward waves increase monotonically with the rotating speed, and those of forward waves decrease initially and then increase gradually with the rotating speed. If only the centrifugal effect is included, which is simulated by omitting the Coriolis terms, $2\Omega\dot{u}$, $2\Omega\dot{v}$, and $2\Omega\dot{w}$ in the dynamic governing equations of the rotating cylindrical shell, the natural frequencies of both the backward and forward waves increase monotonically and there is very little difference between the natural frequencies of the two waves. In this case, we observe that there is a slight difference in the natural frequencies of the two waves at low rotating speeds, but this difference diminishes with increasing rotating speed. These observations indicate that the influences of the Coriolis and centrifugal effects on the frequency characteristics are different and the Coriolis effect is the major factor causing the frequency difference between the backward and forward traveling waves. This is often termed the bifurcation of natural frequencies in rotating shells. In addition, the significance of the rotation influence on the frequency characteristics

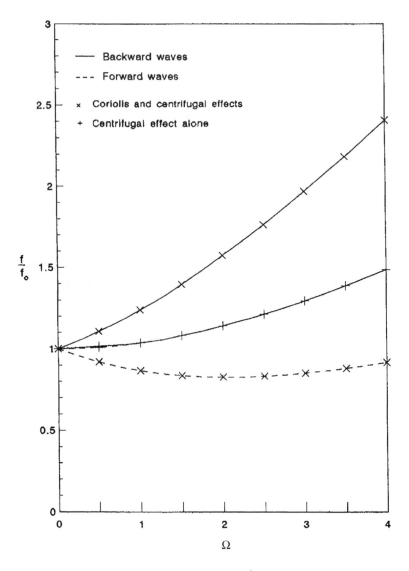

Figure 3.2 Normalized natural frequency f/f_0 as a function of rotating speed Ω rps for a rotating multi-layered cylindrical shell with simply supported boundary condition at both edges ($m = 1$, $n = 2$, $h/R = 0.002$, $L/R = 20$).

is also clearly seen. It can thus be concluded that both the Coriolis and centrifugal effects have significant influences on the frequency characteristics of rotating shells, both qualitatively and quantitatively, and these effects are important and should be considered if rotation is present.

b)   *Different thin-shell theories.*

It is well documented that Love developed the first mathematical framework for a thin-shell theory which is now known as Love's first approximation theory. Following this, the Kirchhoff–Love hypothesis was put forth and has since become the foundation of many thin-shell theories. The developed thin-shell theories using the Kirchhoff–Love hypotheses differ from one another when the terms relevant to h/R are retained or neglected in the constitutive and strain–displacement relations. As such, the governing equations for various thin-shell theories are also expressed in different forms from one another. In this section, a comparative study on the natural frequencies of the forward- and backward-traveling modes for rotating cylindrical shells is carried out to evaluate four well established thin-shell theories — Donnell's, Flügge's, Love's and Sanders' shell theories, by taking a thin multi-layered rotating orthotropic cylindrical shell as an example. This rotating cylindrical shell with the simply supported boundary conditions at both edges is a three equithickness layered orthotropic shell with stacking sequence $[0°/90°/0°]$ and thickness ratio $h/R = 0.002$. The relevant material properties of each layer are $E_{22} = 7.6$ GN/m^2, $E_{11}/E_{22} = 2.5$, $G_{12} = 4.1$ GN/m^2, $\mu_{12} = 0.26$ and $\rho = 1643$ kg/m^3. All numerical comparisons here are for the non-dimensional frequency parameters $f_b^*, f_f^* = \omega R\sqrt{\rho/E_{22}}$ when the axial wavenumber $m = 1$, in which f_b^* and f_f^* denote the backward and forward waves, respectively.

Before discussions on the rotating cylindrical shell, the frequency results of standing-wave modes for stationary cylindrical shells are first studied as a special case for the four thin-shell theories. The non-dimensional frequency parameter $f^* = \omega R\sqrt{\rho/E_{22}}$ for the non-rotating cylindrical shells with $L/R = 1$, 5, 10 and 20 is presented in Table 3.11. It is observed that, in general, the frequency parameters f^* computed by Flügge's, Love's and Sanders' thin-shell theories are in good agreement with each other for most L/R ratios and circumferential wave numbers n. Furthermore, the relative lowest frequency parameters f^* are obtained generally by Sanders' shell theory, followed by Flügge's, Love's and Donnell's thin-shell theories. For Donnell's shell theory, however, the frequency parameters f^* results agree fairly well with those of the rest three shell theories only for small L/R ratios and circumferential wave numbers n. For example, the numerical difference in the frequency results between Donnell's and Sanders' thin-shell theories is about 0.0001% for $L/R = 1$ and $n = 1$, and about 0.03% for $L/R = 1$ and $n = 6$. When the L/R ratios and circumferential wave number n become large, the difference in the frequency parameters f^* increases. For example, the frequency difference between Donnell's and Sanders' thin-shell theories is numerically about 0.02% for $L/R = 20$ and $n = 1$, and about 2.9% for $L/R = 20$ and $n = 6$. It is also seen from the tabulated results in Table 3.11 that, for the non-rotating shells, with the increase of the circumferential wavenumber n, the natural frequencies for all L/R ratios generally

Table 3.11

Comparison of frequency parameter $f^* = \omega R\sqrt{\rho/E_{22}}$ for a $[0°/90°/0°]$ laminated non-rotating cylindrical shell with simply supported boundary conditions at both edges ($m = 1$, $h/R = 0.002$).

L/R	n	Donnell	Flügge	Love	Sanders
1	1	1.061285	1.061283	1.061284	1.061284
	2	0.804058	0.804053	0.804054	0.804053
	3	0.598340	0.598329	0.598331	0.598329
	4	0.450163	0.450141	0.450144	0.450140
	5	0.345288	0.345249	0.345253	0.345247
	6	0.270814	0.270747	0.270754	0.270745
5	1	0.248635	0.248635	0.248635	0.248635
	2	0.107214	0.107203	0.107203	0.107202
	3	0.055140	0.055085	0.055087	0.055085
	4	0.033951	0.033787	0.033790	0.033787
	5	0.026129	0.025790	0.025794	0.025789
	6	0.026362	0.025873	0.025877	0.025871
10	1	0.083910	0.083908	0.083908	0.083908
	2	0.030044	0.030008	0.030009	0.030008
	3	0.015376	0.015191	0.015193	0.015191
	4	0.012605	0.012174	0.012176	0.012173
	5	0.015784	0.015230	0.015231	0.015229
	6	0.021763	0.021178	0.021179	0.021177
20	1	0.023594	0.023590	0.023590	0.023589
	2	0.008032	0.007903	0.007904	0.007903
	3	0.006324	0.005868	0.005869	0.005868
	4	0.009588	0.009019	0.009020	0.009019
	5	0.014823	0.014235	0.014236	0.014235
	6	0.021394	0.020801	0.020801	0.020801

decrease to a minimum frequency before increasing. For example, this minimum frequency occurs at $n = 5$ for $L/R = 5$, $n = 4$ for $L/R = 10$ and $n = 3$ for $L/R = 20$. As such, the value of the circumferential wavenumber n at which the minimum frequency occurs decreases with increasing L/R ratio. For shells with small L/R ratios such as in the case $L/R = 1$, the minimum frequency would occur at a much larger n.

For the thin multi-layered orthotropic rotating cylindrical shell with simply supported boundary conditions and rotating at three speeds $\Omega = 0.1$, 0.4 and 1.0 rps, the non-dimensional frequency parameters f_b^*, $f_f^* = \omega R\sqrt{\rho/E_{22}}$, corresponding to the backward- and forward-traveling modes, respectively, are computed based on the four shell theories — Donnell's, Flügge's, Love's and Sanders'. The numerical frequency results are presented in Tables 3.12–3.21 for the geometric ratios $L/R = 1, 5, 10, 20$, and ∞, respectively.

Table 3.12

Comparison of frequency parameter $f_b^* = \omega R\sqrt{\rho/E_{22}}$ for a $[0°/90°/0°]$ laminated rotating cylindrical shell with simply supported boundary conditions at both edges $(m = 1, h/R = 0.002, L/R = 1)$.

Ω (rps)	n	Donnell	Flügge	Love	Sanders
0.1	1	1.061430	1.061428	1.061429	1.061428
	2	0.804218	0.804213	0.804214	0.804213
	3	0.598485	0.598474	0.598476	0.598473
	4	0.450289	0.450267	0.450270	0.450266
	5	0.345398	0.345358	0.345363	0.345356
	6	0.270912	0.270845	0.270852	0.270843
	7	0.217749	0.217642	0.217651	0.217639
0.4	1	1.016863	1.061861	1.061862	1.061862
	2	0.804700	0.804695	0.804696	0.804695
	3	0.598924	0.598913	0.598915	0.598913
	4	0.450680	0.450659	0.450662	0.450657
	5	0.345759	0.345719	0.345724	0.345718
	6	0.271267	0.271200	0.271207	0.271198
	7	0.218126	0.218020	0.218029	0.218017
1.0	1	1.062729	1.062727	1.062728	1.062727
	2	0.805670	0.805666	0.805667	0.805665
	3	0.599829	0.599818	0.599820	0.599817
	4	0.451532	0.451510	0.451513	0.451509
	5	0.346629	0.346589	0.346593	0.346587
	6	0.272257	0.272190	0.272197	0.272188
	7	0.219366	0.219260	0.219269	0.219257

For comparison of computational accuracy of the four shell theories, the numerical results tabulated in the tables are examined. It is observed that with the use of Flügge's, Love's and Sanders' theories, the computed frequency parameters of the backward- and forward-traveling waves, f_b^* and f_f^*, generally agree well with each other for all the L/R ratios considered. However, the corresponding frequency parameters obtained by Donnell's shell theory are in good agreement with the other three shell theories only when both the L/R ratio and circumferential wave number n are small. As an example, at $\Omega = 0.1$ rps and $n = 7$, the numerical difference of the backward-wave frequency parameters f_b^* between Donnell's and Sanders' thin-shell theories is about 0.05% for the ratio $L/R = 1$, and about 2.1% for the ratio $L/R = 20$. Furthermore, for the four thin-shell theories, the lowest frequency parameters f_b^* and f_f^* of both the backward- and forward-traveling modes are obtained by Sanders' theory, followed by Flügge', Love's and Donnell's shell theories.

Table 3.13

Comparison of frequency parameter $f_f^* = \omega R \sqrt{\rho/E_{22}}$ for a $[0°/90°/0°]$ laminated rotating cylindrical shell with simply supported boundary conditions at both edges ($m = 1$, $h/R = 0.002$, $L/R = 1$).

Ω (rps)	n	Donnell	Flügge	Love	Sanders
0.1	1	1.061141	1.061139	1.061140	1.061139
	2	0.803898	0.803893	0.803894	0.803893
	3	0.598197	0.598186	0.598187	0.598185
	4	0.450040	0.450018	0.450021	0.450017
	5	0.345184	0.345147	0.345149	0.345143
	6	0.270727	0.270660	0.270667	0.270658
	7	0.217587	0.217480	0.217489	0.217477
0.4	1	1.060707	1.060705	1.060706	1.060705
	2	0.803419	0.803414	0.803415	0.803414
	3	0.597771	0.597760	0.597762	0.597760
	4	0.449686	0.449664	0.449667	0.449663
	5	0.344905	0.344865	0.344870	0.344864
	6	0.270528	0.270461	0.270468	0.270459
	7	0.217479	0.217373	0.217382	0.217370
1.0	1	1.059837	1.059835	1.059836	1.059836
	2	0.802468	0.802463	0.802464	0.802463
	3	0.596946	0.596935	0.596937	0.596934
	4	0.449046	0.449024	0.449027	0.449023
	5	0.344494	0.344454	0.344459	0.344452
	6	0.270410	0.270343	0.270349	0.270340
	7	0.217749	0.217642	0.217651	0.217639

The rotating speed is found to have some influence on the numerical difference between the predicted frequency parameters f_b^* and f_f^* by the different shell theories. As an illustration, for the ratio $L/R = 5$ and $n = 7$, the difference in the predicted frequency results between Donnell's and Sanders' theories is about 1.8% at $\Omega = 0.1$ rps, and about 1.3% at $\Omega = 1.0$ rps. When $L/R = 20$ and $n = 7$, this frequency difference becomes about 2.1% at $\Omega = 0.1$ rps, and about 1.4% at $\Omega = 1.0$ rps. It is shown that, arising from the rotating speed Ω, the discrepancy of the computed frequency parameters f_b^* and f_f^* from the different thin-shell theories increases with the L/R ratio. With the increase in the rotating speed Ω, however, the Donnell frequency parameters agree better with those of the other three thin-shell theories.

Certain minimum frequency parameters are observed to occur for both the backward- and forward-traveling waves, with the increase of the circumferential wave number n. It appears that the frequencies f_b^* and f_f^* generally decreases with n until the minimum frequencies are reached, upon which they will subsequently increase with n. For a rotating cylindrical shell with larger L/R ratio, the minimum frequency parameters of both the traveling waves usually occur at lower circumferential wave number n.

Table 3.14

Comparison of frequency parameter $f_b^* = \omega R\sqrt{\rho/E_{22}}$ for a $[0°/90°/0°]$ laminated rotating cylindrical shell with simply supported boundary conditions at both edges ($m = 1$, $h/R = 0.002$, $L/R = 5$).

Ω (rps)	n	Donnell	Flügge	Love	Sanders
0.1	1	0.248918	0.248917	0.248917	0.248917
	2	0.107447	0.107435	0.107436	0.107435
	3	0.055320	0.055266	0.055267	0.055265
	4	0.034105	0.033942	0.033945	0.033941
	5	0.026277	0.025939	0.025943	0.025938
	6	0.026510	0.026022	0.026026	0.026021
	7	0.031645	0.031086	0.031089	0.031085
0.4	1	0.249765	0.249765	0.249765	0.249764
	2	0.108154	0.108143	0.108143	0.108143
	3	0.055921	0.055867	0.055868	0.055967
	4	0.034768	0.034606	0.034608	0.034605
	5	0.027152	0.026821	0.026825	0.026820
	6	0.027586	0.027112	0.027116	0.027111
	7	0.032812	0.032269	0.032272	0.032268
1.0	1	0.251458	0.251458	0.251458	0.251458
	2	0.109610	0.109599	0.109600	0.109599
	3	0.057393	0.057339	0.057341	0.057339
	4	0.036960	0.036803	0.036806	0.036803
	5	0.030657	0.030357	0.030361	0.030356
	6	0.032218	0.031806	0.031810	0.031805
	7	0.038044	0.037571	0.037574	0.037571

For example, the minimum frequency parameters of the traveling waves occur at $n = 5$ if the ratio $L/R = 5$, and occur at $n = 4$ if the ratio $L/R = 10$. It also appears that the value of the circumferential wave number n at which the minimum frequency parameters occur is dependent on the rotational speed Ω. For example, when $L/R = 20$, the minimum frequency parameters occur at $n = 3$ for the rotating speeds $\Omega = 0.1$ and 0.4 rps, and at $n = 2$ for the rotating speed $\Omega = 1.0$ rps.

In general, the frequency parameters f_b^* of the backward modes increase with the rotating speed Ω for all the circumferential wavenumbers n. For the forward-traveling modes, the frequency parameters f_f^* at certain circumferential wavenumbers n first decrease and then increase with the rotating speed Ω, and at other wavenumbers n the parameters f_f^* increase monotonically with increasing Ω. This characteristic of frequency parameters f_f^* of the forward modes is dependent on the L/R ratio. For example, when $L/R = 5$ and circumferential wave numbers $n \geq 5$, the frequency parameters f_f^* of forward modes increase monotonically with the rotating speed Ω. However, when

Table 3.15

Comparison of frequency parameter $f_f^* = \omega R\sqrt{\rho/E_{22}}$ for a $[0°/90°/0°]$ laminated rotating cylindrical shell with simply supported boundary conditions at both edges ($m = 1$, $h/R = 0.002$, $L/R = 5$).

Ω (rps)	n	Donnell	Flügge	Love	Sanders
0.1	1	0.248353	0.248352	0.248352	0.248352
	2	0.106983	0.106972	0.106972	0.106971
	3	0.054969	0.054915	0.054916	0.054914
	4	0.033830	0.033666	0.033669	0.003366
	5	0.026052	0.025714	0.025718	0.025713
	6	0.026321	0.025832	0.025836	0.025831
	7	0.031481	0.030922	0.030925	0.030921
0.4	1	0.247504	0.247504	0.247504	0.247504
	2	0.106298	0.106287	0.106288	0.106287
	3	0.054519	0.054464	0.054466	0.054464
	4	0.033666	0.033504	0.033507	0.033503
	5	0.026252	0.025920	0.025924	0.025919
	6	0.026827	0.026353	0.026357	0.026352
	7	0.032157	0.031614	0.031617	0.031613
1.0	1	0.245807	0.245806	0.245806	0.245806
	2	0.104971	0.104960	0.104961	0.104960
	3	0.053886	0.053832	0.053834	0.053832
	4	0.034205	0.034049	0.034052	0.034049
	5	0.028406	0.028106	0.028110	0.028106
	6	0.030320	0.029909	0.029912	0.029908
	7	0.036406	0.035934	0.035936	0.035933

$L/R = 20$, the frequency parameters f_f^* increase monotonically with the rotating speed Ω for circumferential wave numbers $n \geq 3$.

In conclusion, Sanders' shell theory provides the most accurate frequency results of the four thin-shell theories. In terms of complexities, Flügge's shell theory is the most complicated due to its more complex kinematic strain–displacement and constitutive relations. Donnell's shell theory is the simplest of the four theories, and the numerical results indicate that Donnell's shell theory is sufficiently accurate for rotating cylindrical shells with small L/R ratios and when the circumferential wave number n is small. In terms of both simplicity and accuracy, Love's shell theory is the best among the four thin-shell theories, especially when the L/R ratios are high or the circumferential wave number n is not low.

c)   *Influence of rotating velocity.*

The effects of rotating angular velocity in rotating shells constitute one of the most critical investigations of this monograph. Physically, the important differences between

Table 3.16

Comparison of frequency parameter $f_b^* = \omega R\sqrt{\rho/E_{22}}$ for a $[0°/90°/0°]$ laminated rotating cylindrical shell with simply supported boundary conditions at both edges ($m = 1$, $h/R = 0.002$, $L/R = 10$).

Ω (rps)	n	Donnell	Flügge	Love	Sanders
0.1	1	0.084194	0.084193	0.084193	0.084193
	2	0.030280	0.030244	0.030245	0.030244
	3	0.015570	0.015385	0.015387	0.015385
	4	0.012790	0.012358	0.012360	0.012357
	5	0.015956	0.015404	0.015406	0.015404
	6	0.021923	0.021339	0.021341	0.021339
	7	0.029527	0.028934	0.028935	0.028934
0.4	1	0.085047	0.085046	0.085046	0.085046
	2	0.031022	0.030987	0.030987	0.030987
	3	0.016366	0.016184	0.016185	0.016183
	4	0.013857	0.013450	0.013451	0.013449
	5	0.017167	0.016645	0.016646	0.016644
	6	0.023159	0.022600	0.022601	0.022600
	7	0.030758	0.030185	0.030186	0.030185
1.0	1	0.086750	0.086749	0.086749	0.086749
	2	0.032659	0.032624	0.032625	0.032624
	3	0.018849	0.018683	0.018684	0.018682
	4	0.017887	0.017560	0.017561	0.017559
	5	0.022070	0.021656	0.021657	0.021656
	6	0.028454	0.027993	0.027994	0.027993
	7	0.036267	0.035777	0.035777	0.035776

the rotating and non-rotating shells of revolution are the Coriolis and centrifugal accelerations, as well as the hoop tension arising in rotating shells due to the angular velocities. These effects have significant influences on the dynamic behaviors of the rotating shells. For example, the frequency characteristics of a stationary shell structure are generally determined by the shell geometry, material properties and boundary conditions. However, when the shell rotates, the structural frequency characteristics are qualitatively altered. This qualitative difference manifests itself in the form of a bifurcation phenomenon in the natural frequency parameters. For a stationary shell of revolution, the vibration of the shell is a standing wave motion. However, when the same shell rotates, the standing wave motion is transformed, and depending on the direction of rotation, backward or forward waves will emerge. In this section, the discussion is made for the influence of the rotating velocity on the frequency characteristics of the rotating circular cylindrical shell.

Figure 3.3 shows the variation of the natural frequencies f with the circumferential wave numbers n. The figure illustrates the effect of rotation on the frequencies of the rotating multi-layered cylindrical shell which has three isotropic

Table 3.17

Comparison of frequency parameter $f_f^* = \omega R\sqrt{\rho/E_{22}}$ for a $[0°/90°/0°]$ laminated rotating cylindrical shell with simply supported boundary conditions at both edges ($m = 1$, $h/R = 0.002$, $L/R = 10$).

Ω (rps)	n	Donnell	Flügge	Love	Sanders
0.1	1	0.083625	0.083624	0.083624	0.083624
	2	0.029813	0.029778	0.029778	0.029778
	3	0.015219	0.015034	0.015036	0.015034
	4	0.012512	0.012083	0.012085	0.012082
	5	0.015731	0.015179	0.015181	0.015179
	6	0.021733	0.021150	0.021151	0.021149
	7	0.029363	0.028771	0.028772	0.028770
0.4	1	0.082771	0.082770	0.082770	0.082770
	2	0.029156	0.029121	0.029122	0.029121
	3	0.014963	0.014781	0.014783	0.014781
	4	0.012757	0.012349	0.012351	0.012349
	5	0.016268	0.015745	0.015747	0.015745
	6	0.022400	0.021842	0.021843	0.021841
	7	0.030103	0.029531	0.029532	0.029530
1.0	1	0.081061	0.081060	0.081060	0.081060
	2	0.027995	0.027960	0.027961	0.027960
	3	0.015342	0.015176	0.015177	0.015175
	4	0.015136	0.014809	0.014810	0.014808
	5	0.019822	0.019407	0.019409	0.019407
	6	0.026558	0.026097	0.026098	0.026097
	7	0.034631	0.034140	0.034141	0.034140

layers of construction, where the thickness of middle layer ($E = 2.0685 \times 10^{11}$ N/m², $\mu = 0.3$ and $\rho = 8053$ kg/m³) is three times that of the two surface layers ($E = 4.8265 \times 10^9$ N/m², $\mu = 0.3$ and $\rho = 1314$ kg/m³). In the figure the frequency curves for $\Omega = 0$, 1 and 4 rps are shown. It is observed that the general frequency characteristics of a rotating shell are similar to those of a stationary shell. The natural frequencies of both the backward and forward waves first decrease with the wave number and then increase. Before the fundamental frequency is reached, the natural frequencies of the backward waves are larger than those of a stationary shell, and for the forward waves the natural frequencies are lower than those of the stationary shell. All the frequencies of both backward and forward waves are always larger than those of the stationary shell after the fundamental frequency is attained. For any rotating velocity, the difference between the natural frequencies of the forward and backward waves is always greater for a smaller circumferential wave number, and this difference diminishes as the circumferential wave number increases. At low rotating velocities the difference

Table 3.18

Comparison of frequency parameter $f_b^* = \omega R\sqrt{\rho/E_{22}}$ for a $[0°/90°/0°]$ laminated rotating cylindrical shell with simply supported boundary conditions at both edges ($m = 1$, $h/R = 0.002$, $L/R = 20$).

Ω (rps)	n	Donnell	Flügge	Love	Sanders
0.1	1	0.023883	0.023879	0.023879	0.023879
	2	0.008276	0.008148	0.008148	0.008147
	3	0.006544	0.006091	0.006092	0.006090
	4	0.009785	0.009219	0.009220	0.009219
	5	0.014999	0.014414	0.014414	0.014414
	6	0.021554	0.020963	0.020964	0.020963
	7	0.029346	0.028752	0.028752	0.028752
0.4	1	0.024749	0.024745	0.024745	0.024745
	2	0.009137	0.009011	0.009012	0.009011
	3	0.007697	0.007288	0.007289	0.007287
	4	0.011045	0.010528	0.010528	0.010527
	5	0.016263	0.015714	0.015714	0.015713
	6	0.022806	0.022241	0.022241	0.022241
	7	0.030583	0.030009	0.030009	0.030009
1.0	1	0.026480	0.026476	0.026476	0.026476
	2	0.011382	0.011268	0.011269	0.011268
	3	0.011548	0.011260	0.011260	0.011259
	4	0.015719	0.015345	0.015345	0.015345
	5	0.021357	0.020931	0.020931	0.020930
	6	0.028163	0.027699	0.027700	0.027699
	7	0.036118	0.035627	0.035627	0.035627

between the natural frequencies of the two waves is small compared with the corresponding case when the rotating velocity is high. This indicates that the rotational effect is significant at high rotational speeds.

For a laminated composite cylindrical shell, Fig. 3.4 shows the variation of the fundamental frequencies f_f (Hz) with the rotating velocity Ω for a cylindrical shell with simply supported boundary condition at both edges and $h/R = 0.002$, $L/R = 20$. The cylindrical shell is a three equithickness layered orthotropic shell with a stacking sequence $[90°/0°/90°]$, and the relevant material properties of each layer are $E_{22} = 7.6 \text{ GN/m}^2$, $E_{11}/E_{22} = 2.5$, $G_{12} = 4.1 \text{ GN/m}^2$, $\mu_{12} = 0.26$, $\rho = 1643 \text{ kg/m}^3$. From the figure, the fundamental frequencies for the backward-waves increase monotonically with the rotating velocity while those for the forward-waves first decrease and then increase gradually with the rotating velocity. The figure also shows that the fundamental frequencies for the backward and forward waves at any rotating velocity are different and those for the backward-waves are always higher than those for the forward-waves, which is the result of the Coriolis influence.

Table 3.19

Comparison of frequency parameter $f_f^* = \omega R\sqrt{\rho/E_{22}}$ for a $[0°/90°/0°]$ laminated rotating cylindrical shell with simply supported boundary conditions at both edges ($m = 1$, $h/R = 0.002$, $L/R = 20$).

Ω (rps)	n	Donnell	Flügge	Love	Sanders
0.1	1	0.023304	0.023300	0.023300	0.023300
	2	0.007809	0.007681	0.007681	0.007680
	3	0.006193	0.005740	0.005741	0.005739
	4	0.009510	0.008944	0.008945	0.008944
	5	0.014775	0.014190	0.014190	0.014189
	6	0.021365	0.020774	0.020774	0.020774
	7	0.029182	0.028588	0.028589	0.028588
0.4	1	0.022436	0.022432	0.022432	0.022432
	2	0.007268	0.007142	0.007143	0.007142
	3	0.006295	0.005885	0.005886	0.005885
	4	0.009945	0.009428	0.009428	0.009427
	5	0.015364	0.014815	0.014815	0.014815
	6	0.022048	0.021483	0.021483	0.021483
	7	0.029929	0.029355	0.029355	0.029355
1.0	1	0.020698	0.020694	0.020694	0.020694
	2	0.006711	0.006597	0.006597	0.006597
	3	0.008042	0.007753	0.007754	0.007753
	4	0.012969	0.012595	0.012595	0.012595
	5	0.019109	0.018683	0.018683	0.018683
	6	0.026268	0.025804	0.025804	0.025804
	7	0.034481	0.033990	0.033991	0.033990

For the same laminated composite orthotropic cylindrical shell as in Fig. 3.4, Fig. 3.5 shows the variation of the non-dimensional frequency parameter $\omega^* = \omega L_s\sqrt{\rho h/A_{11}}$ (ω (rad/s) is the natural circular frequency, $L_s = 2R\pi$, R is the radius, and ρ and A_{11} are the density and extensional stiffness) with the circumferential wave number n at three rotating velocities $\Omega = 1$, 2 and 3 rps. From the figure, the general behavior of the frequency curves is to first decrease to a minima and then increase with the circumferential wave number n. This behavior is similar to that for a non-rotating cylindrical shell.

Figure 3.6 shows the variation of the fundamental frequencies f_f (Hz) with the thickness-to-radius ratio h/R for the same laminated composite orthotropic cylindrical shell used in Fig. 3.4, at three rotating velocities $\Omega = 1$, 2 and 3 rps. The fundamental frequencies for both the backward and forward waves at any rotating velocity are found to increase monotonically with the h/R ratio. The fundamental frequencies for the backward waves increase monotonically with rotating velocity at any h/R ratio. However, for the forward waves, this trend is only observed at small h/R ratios. At higher

Table 3.20

Comparison of frequency parameter $f_b^* = \omega R\sqrt{\rho/E_{22}}$ for a laminated $[0°/90°/0°]$ infinitely long rotating cylindrical shell ($m = 1$, $h/R = 0.002$).

Ω (rps)	n	Donnell	Flügge	Love	Sanders
0.1	1	0.000717	0.000292	0.000292	0.000292
	2	0.002425	0.001900	0.001900	0.001900
	3	0.005364	0.004800	0.004800	0.004800
	4	0.009533	0.008953	0.008953	0.008953
	5	0.014919	0.014332	0.014332	0.014332
	6	0.021515	0.020924	0.020924	0.020924
	7	0.029320	0.028726	0.028726	0.028726
0.4	1	0.001594	0.001169	0.001169	0.001169
	2	0.003658	0.003256	0.003256	0.003256
	3	0.006644	0.006159	0.006159	0.006159
	4	0.010814	0.010286	0.010286	0.010286
	5	0.016187	0.015636	0.015636	0.015636
	6	0.022768	0.022203	0.022203	0.022203
	7	0.030558	0.029984	0.029984	0.029984
1.0	1	0.003347	0.002921	0.002921	0.002921
	2	0.007032	0.006811	0.006811	0.006811
	3	0.010828	0.010517	0.010517	0.010517
	4	0.015551	0.015174	0.015174	0.015174
	5	0.021298	0.020871	0.020871	0.020871
	6	0.028132	0.027669	0.027669	0.027699
	7	0.036096	0.035605	0.035605	0.035605

h/R ratios, the fundamental frequencies are observed to decrease with the rotating velocity.

Figure 3.7 shows the variation of non-dimensional fundamental frequency parameter $\omega_f^* = \omega_f L_s\sqrt{\rho h/A_{11}}$ (ω_f (rad/s) is the fundamental circular frequency, $L_s = 2R\pi$, R is the radius, and ρ and A_{11} are the density and extensional stiffness term) with the length-to-radius ratio L/R for the same laminated composite orthotropic cylindrical shell used in Fig. 3.4 at three rotating velocities $\Omega = 1$, 2 and 3 rps. The fundamental frequencies for the backward and forward waves decrease with the L/R ratio for all rotating velocities, and those with high rotating velocities are generally higher than those with low rotating velocities.

Figures 3.8–3.10 show the relative displacements u, v, and w along the x-direction for the same rotating multi-layered cylindrical shell used in Fig. 3.3, with three isotropic layers of construction and rotating at velocity $\Omega = 1$ rps. It is found that rotation has significant influence on the vibrational displacement responses and it is most significant at the location where the displacements for a stationary cylindrical shell are at the maximum. Further

Table 3.21

Comparison of frequency parameter $f_f^* = \omega R\sqrt{\rho/E_{22}}$ for a laminated $[0°/90°/0°]$ infinitely long rotating cylindrical shell ($m = 1$, $h/R = 0.002$).

Ω (rps)	n	Donnell	Flügge	Love	Sanders
0.1	1	0.000133	0.000292	0.000292	0.000292
	2	0.001958	0.001433	0.001433	0.001433
	3	0.005013	0.004450	0.004450	0.004450
	4	0.009258	0.008678	0.008678	0.008678
	5	0.014694	0.014107	0.014107	0.014107
	6	0.021326	0.020734	0.020734	0.020734
	7	0.029156	0.028562	0.028562	0.028562
0.4	1	0.000743	0.001168	0.001169	0.001169
	2	0.001788	0.001387	0.001387	0.001387
	3	0.005242	0.004757	0.004757	0.004757
	4	0.009714	0.009186	0.009186	0.009186
	5	0.015288	0.014737	0.014737	0.014737
	6	0.022010	0.021445	0.021445	0.021445
	7	0.029903	0.029329	0.029329	0.029329
1.0	1	0.002496	0.002921	0.002921	0.002921
	2	0.002357	0.002137	0.002137	0.002137
	3	0.007322	0.007012	0.007012	0.007012
	4	0.012802	0.012425	0.012425	0.012425
	5	0.019051	0.018624	0.018624	0.018624
	6	0.026237	0.025774	0.025774	0.025774
	7	0.034460	0.033969	0.033969	0.033969

examination also shows that the difference in the relative displacements between a stationary cylindrical shell $\Omega = 0$ and those due to backward and forward waves at $\Omega = 1$ rps in the u-direction are 6.4 and 14.3%, respectively. In the v-direction, the corresponding differences are 82 and 71.2%, respectively, and in the w-direction they are 6.4 and 14.2%, respectively. This indicates that the influence of rotational motion on the vibrational displacement responses of cylindrical shell is significant and is also most pronounced in the v-direction.

The discussions made in this section show that rotational motion has significant influence on both the natural frequencies and vibrational displacements and these influences depend largely on the velocity of rotation.

d)   *Influence of length and thickness.*

There are generally many physical and geometrical parameters which influence the frequency characteristics of rotating shells. Physical parameters include the rotating angular velocity, material properties and boundary conditions. The major geometrical

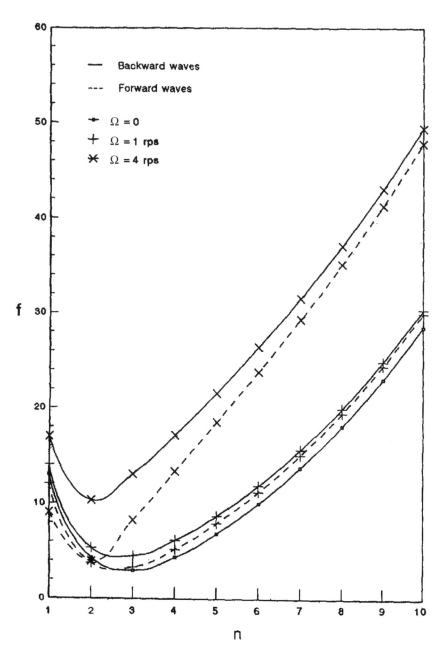

Figure 3.3 Natural frequency f (Hz) as a function of circumferential wave numbers n for a rotating multi-layered cylindrical shell with simply supported boundary condition at both edges ($m = 1$, $h/R = 0.002$, $L/R = 20$).

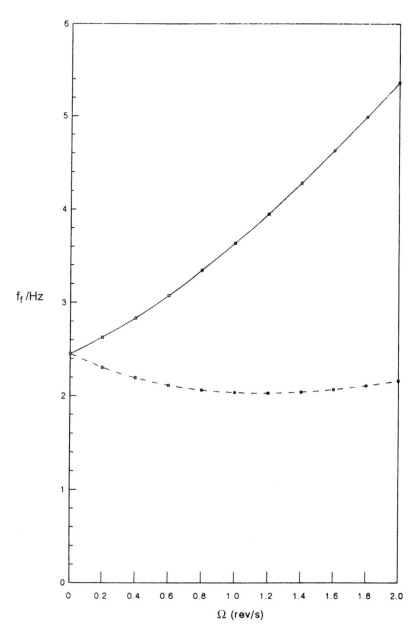

Figure 3.4 Variation of fundamental frequencies f_f (Hz) with the rotating velocity Ω for a rotating laminated composite orthotropic cylindrical shell with simply supported boundary condition at both edges ($h/R = 0.002$, $L/R = 20$).

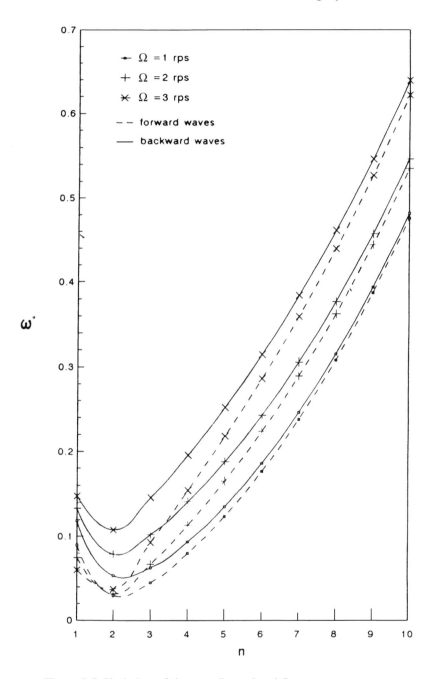

Figure 3.5 Variation of the non-dimensional frequency parameter
$\omega^* = \omega L_s\sqrt{\rho h / A_{11}}$ with the circumferential wave number n for a rotating
laminated composite orthotropic cylindrical shell with simply supported
boundary condition at both edges ($m = 1$, $h/R = 0.002$, $L/R = 20$).

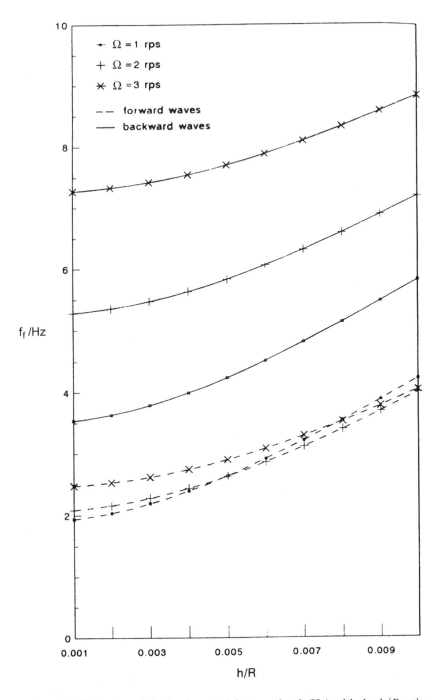

Figure 3.6 Variation of the fundamental frequencies f_f (Hz) with the h/R ratio for a rotating laminated composite orthotropic cylindrical shell with simply supported boundary condition at both edges ($m = 1$, $L/R = 20$).

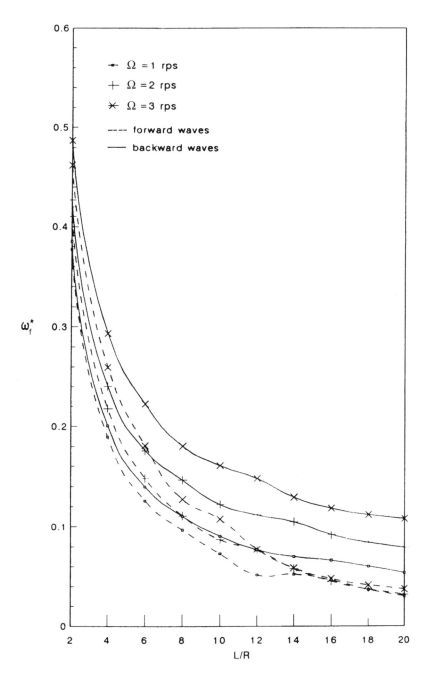

Figure 3.7 Variation of non-dimensional fundamental frequency parameter $\omega_f^* = \omega_f L_s \sqrt{\rho h / A_{11}}$ with the L/R ratio for a rotating laminated composite orthotropic cylindrical shell with simply supported boundary condition at both edges ($m = 1$, $h/R = 0.002$).

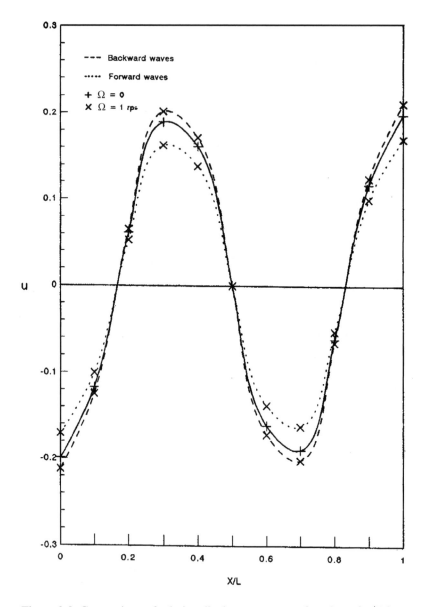

Figure 3.8 Comparison of relative displacement u as a function of x/L for a rotating multi-layered cylindrical shell with simply supported boundary condition at both edges ($m = 1$, $n = 1$, $h/R = 0.002$, $L/R = 20$).

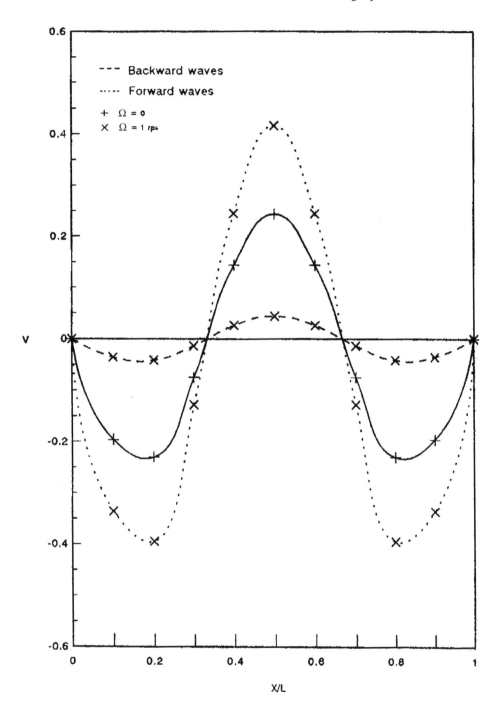

Figure 3.9 Comparison of relative displacement v as a function of x/L for a rotating multi-layered cylindrical shell with simply supported boundary condition at both edges ($m = 1$, $n = 1$, $h/R = 0.002$, $L/R = 20$).

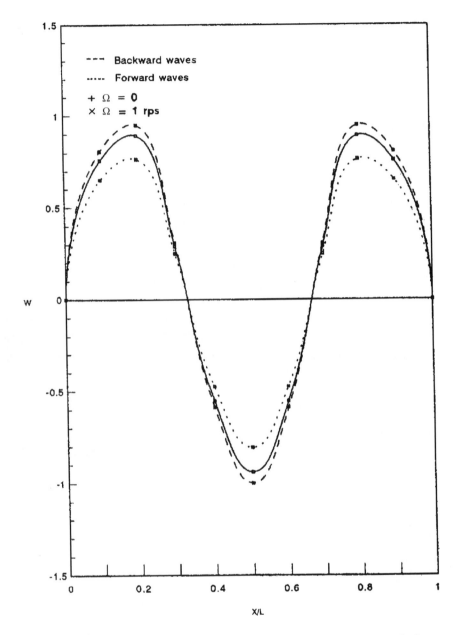

Figure 3.10 Comparison of relative displacement w as a function of x/L for a rotating multi-layered cylindrical shell with simply supported boundary condition at both edges ($m = 1$, $n = 1$, $h/R = 0.002$, $L/R = 20$).

parameters include the length L, radius R and thickness h. In this section, discussions are made on the influence of the geometrical length ratio L/R and thickness ratio h/R on the frequency characteristics of the rotating multi-layered circular cylindrical shell. The presently discussed cylindrical shell has three isotropic layers of construction, where the thickness of the middle layer ($E = 2.0685 \times 10^{11}$ N/m^2, $\mu = 0.3$ and $\rho = 8053$ kg/m^3) is three times as thick as those of the two surface layers ($E = 4.8265 \times 10^9$ N/m^2, $\mu = 0.3$ and $\rho = 1314$ kg/m^3).

Figure 3.11 shows the variation of the natural frequencies f (Hz) with the rotating velocity Ω (rps) at various thickness ratios h/R for the rotating multi-layered cylindrical shell with the simply supported boundary conditions at both edges. It is clearly seen that the natural frequencies for both the forward and backward waves increase with the thickness ratio h/R at all the rotational velocities Ω. With the increase of rotating velocity Ω, however, the natural frequencies of the forward wave initially decrease while those of the backward wave generally increase at all h/R values.

Figure 3.12 shows the variation of the natural frequencies f (Hz) with the rotating velocity Ω (rps) at various length ratios L/R for the rotating multi-layered cylindrical shell with the simply supported boundary conditions at both edges. It is observed that the natural frequencies for both the backward and forward waves generally decrease with the length ratio L/R at all the rotational velocities Ω. With increasing the rotating velocity Ω, the natural frequencies of the forward wave initially decrease while those of the backward wave generally increase at all L/R values. It is further observed that the influence of the length ratio L/R on the natural frequency of the rotating cylindrical shell is larger than that of the thickness ratio h/R.

Variations of the fundamental frequencies f_f (Hz) with the length ratio L/R at different rotating velocities are shown in Fig. 3.13 ($\Omega = 0.5$ rps) and Fig. 3.14 ($\Omega = 4$ rps) for various thickness ratios h/R of the rotating multi-layered cylindrical shell with the simply supported boundary conditions at both edges. In both the figures, the fundamental frequencies f_f (Hz) for the two waves decrease with the L/R ratio. The fundamental frequencies f_f of the backward and forward waves of a thicker shell are always larger than those of the corresponding waves of a thinner one. For the cylindrical shell rotating at lower velocity $\Omega = 0.5$ rps, bifurcation of fundamental frequencies for small L/R ratio is not significant. The bifurcation of fundamental frequencies f_f (Hz) becomes more significant as the L/R ratio becomes larger. However, for the cylindrical shell rotating at higher velocity $\Omega = 4$ rps, the bifurcation of fundamental frequencies is relatively more significant at small L/R ratios. It also appears that the thickness ratio h/R has little or no effect on the bifurcation characteristics of the fundamental frequencies f_f. This may be noted from the apparently similar difference in fundamental frequencies between the backward and forward waves at $h/R = 0.002$ and 0.01 for both the velocities $\Omega = 0.5$ and 4 rps.

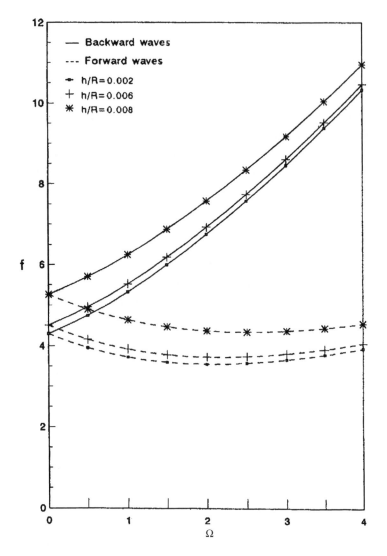

Figure 3.11 Natural frequency f (Hz) as a function of rotating
velocity Ω rps at different h/R ratios for a simply supported rotating
multi-layered cylindrical shell ($m = 1$, $n = 2$, $L/R = 20$).

e)   ***Influence of layered configuration of composites.***

When the rotating shell is made of composite material, the influence of the
layered configuration should be considered since it is one of the most important
characteristics of a composite material. Usually layers are made of different orthotropic
materials, and their principal directions may also be oriented differently. For laminated

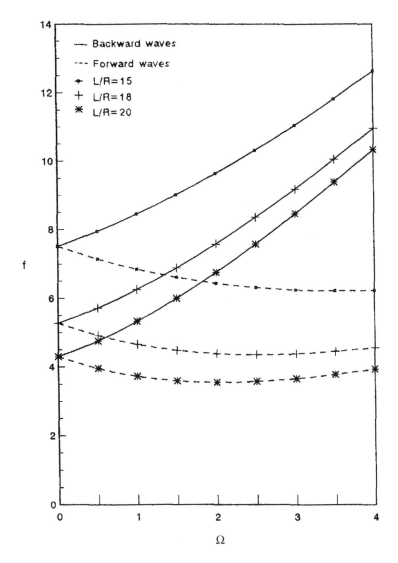

Figure 3.12 Natural frequency f (Hz) as a function of rotating velocity Ω (rps) at different L/R ratios for a simply supported rotating multi-layered cylindrical shell ($m = 1, n = 2, h/R = 0.002$).

composites, the fiber directions determine layer orientation. In this section, however, the discussion is made for a simplified case where the rotating cylindrical shells are composed of three laminated isotropic layers, called the inner, middle and outer layers, with different thicknesses. The material properties of both the inner and outer layers are taken as $E = 4.8265 \times 10^9 \, \text{N/m}^2$, $\mu = 0.3$ and $\rho = 1314 \, \text{kg/m}^3$, and those

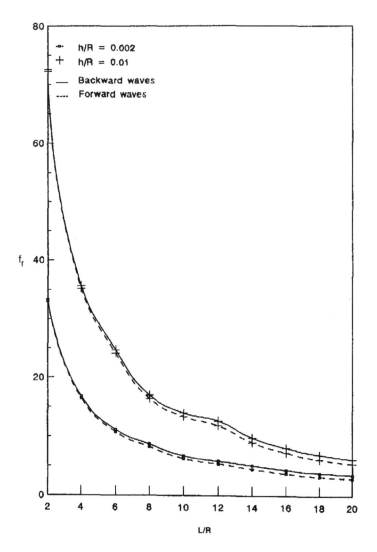

Figure 3.13 Fundamental frequency f_f (Hz) as a function of the L/R ratio at different h/R ratios for a simply supported rotating multi-layered cylindrical shell ($m = 1$, $n = 2$, $\Omega = 0.5$ rps).

of the middle layer are $E = 2.0685 \times 10^{11}$ N/m^2, $\mu = 0.3$ and $\rho = 8053$ kg/m^3. Three rotating cylindrical shells studied here are termed Cylinders I, II and III for identification purposes. Cylinder I has a thicker middle layer that is three times as thick as the thicknesses of both the inner and outer layers, Cylinder II has a thinner middle layer that is three times as thin as those of the other two layers, and Cylinder III has equal thicknesses for all the three layers.

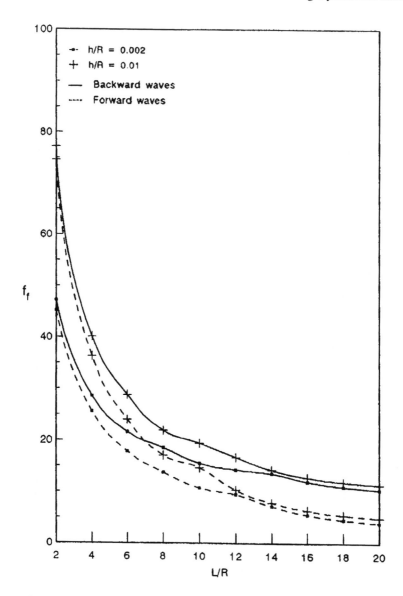

Figure 3.14 Fundamental frequency f_f (Hz) as a function of the L/R ratio at different h/R ratios for a simply supported rotating multi-layered cylindrical shell ($m = 1$, $n = 2$, $\Omega = 4$ rps).

Figures 3.15–3.17 show the influence of the layered configuration on the natural frequencies of the rotating cylindrical shells with simply supported boundary condition at both edges.

Figure 3.15 shows the variation of the natural frequencies f (Hz) with the rotating velocity Ω for Cylinder I, Cylinder II and Cylinder III. It is observed that Cylinder I has

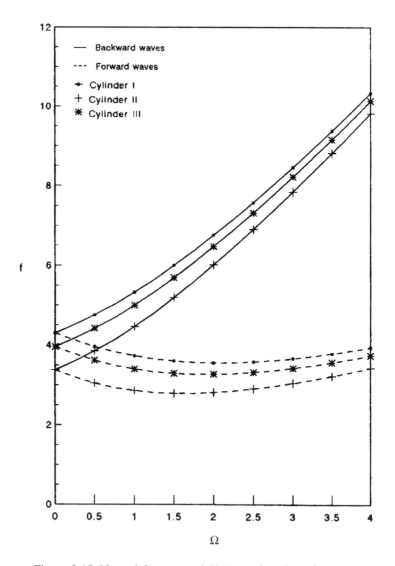

Figure 3.15 Natural frequency f (Hz) as a function of rotating velocity Ω for simply supported rotating cylindrical shells with different layered configurations ($m = 1$, $n = 2$, $h/R = 0.002$, $L/R = 20$).

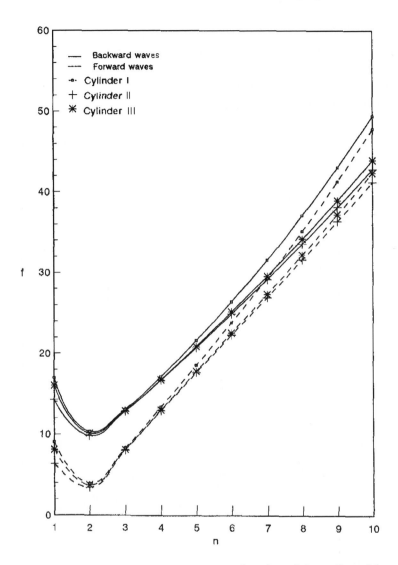

Figure 3.16 Natural frequency f (Hz) as a function of circumferential wave numbers n for simply supported rotating cylindrical shells with different layered configurations ($m = 1$, $n = 2$, $h/R = 0.002$, $L/R = 20$).

the highest natural frequencies for both the forward and backward waves, followed by Cylinder III, and the natural frequencies of Cylinder II are relatively the lowest.

Figure 3.16 shows the variation of the natural frequencies f (Hz) with the circumferential wave numbers n for Cylinder I, Cylinder II and Cylinder III. The natural

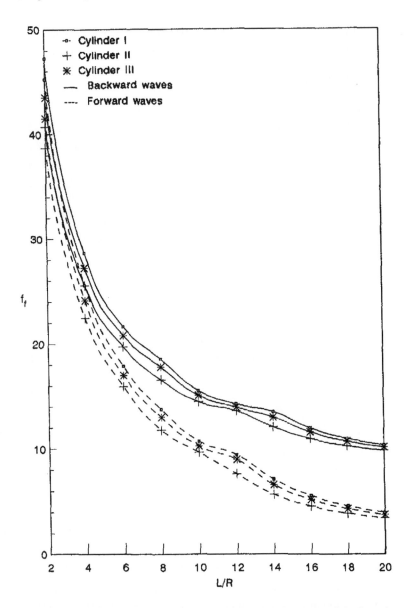

Figure 3.17 Fundamental frequencies f_f (Hz) as a function of the length ratio L/R for simply supported rotating cylindrical shells with different layered configurations ($\Omega = 4$ rps, $h/R = 0.002$).

frequencies corresponding to the forward and backward waves for the three cylindrical shells first decrease and then increase with the circumferential wave numbers n. Cylinder I possesses the relative highest natural frequencies, followed by Cylinder III and Cylinder II. It is also evident that, with the increase of the circumferential wave number n, the differences between the natural frequencies associated with the forward and backward waves gradually decrease for all the three cylindrical shells considered here.

Figure 3.17 shows the variation of the fundamental frequencies f_f (Hz) with the length ratio L/R for various layered configurations. As the length ratio L/R is increased for the three cylindrical shells, the fundamental frequencies for both the forward and backward waves initially decrease rapidly and this is subsequently followed by a more gradual decrease. Once again, Cylinder I has the relative highest fundamental frequencies, followed by Cylinder III and Cylinder II.

f)   *Influence of boundary condition.*

In generic boundary value problems of engineering shell structures, most of the common boundary conditions fall into two categories: in-reference (middle) surface and bending. The in-reference-surface boundary conditions on the loaded edges are known from the applied loading. On an unloaded edge, many possibilities exist for the in-reference-surface boundary conditions. For example, an unloaded edge free to deform in the reference surface is referred to as a moveable edge; or the movement of the unloaded edge may be restrained in the direction normal to the edge but is free to expand or contract longitudinally, in which case the boundary condition is referred to as an immovable edge. On the other hand, typical bending boundary conditions include the free boundary condition that is free to deform in any direction, the clamped boundary condition where both the out-of-reference-surface displacement and rotation about the edge are rigidly fixed, and the simply supported boundary condition with out-of-reference-surface rotation about the edge allowed while the out-of-reference-surface displacement is restrained.

In all studies herein for the rotating cylindrical shells, four boundary conditions are considered. They are the simply supported boundary condition at both edges (SS–SS), the clamped boundary condition at both edges (C–C), the clamped–simply supported boundary condition (C–SS) and the clamped–free boundary condition (C–F). Additionally, the rotating cylindrical shell studied here is a thin laminated composite shell, consisting of three orthotropic equithickness layers with a stacking sequence [90°/0°/90°]. The relevant material properties of each layer are $E_{22} = 7.6\ \mathrm{GN/m^2}$, $E_{11}/E_{22} = 2.5$, $G_{12} = 4.1\ \mathrm{GN/m^2}$, $\mu_{12} = 0.26$, $\rho = 1643\ \mathrm{kg/m^3}$.

In the present study of the influence of boundary condition on the dynamic characteristics of the rotating cylindrical shells, four investigations are made on the frequency parameter $f^* = fL_s\sqrt{\rho h/A_{11}}$ (f (Hz) is the natural frequency, $L_s = 2R\pi$, R is the radius, ρ and A_{11} are the density and extensional stiffness term of the rotating shell). The first is to study how the frequency parameter f^* varies with the circumferential wave number n for different rotating velocities Ω. The results are presented in Figs. 3.18–3.20. The second is to study the variation of the frequency parameter f^* with the rotating velocities Ω for different length ratios L/R and thickness ratios h/R. The results are presented in Figs. 3.21–3.24. The third is to study the variation of the frequency parameter f^* with the geometric properties L/R and h/R of the rotating cylindrical shell. The results are presented in Figs. 3.25–3.26. The fourth is to study the influence of

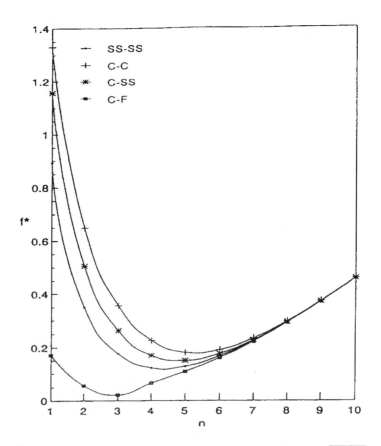

Figure 3.18 Variation of the frequency parameter $f^* = fL_s\sqrt{\rho h/A_{11}}$ with the circumferential wave number n ($m = 1$, $h/R = 0.002$, $L/R = 6$, $\Omega = 0$).

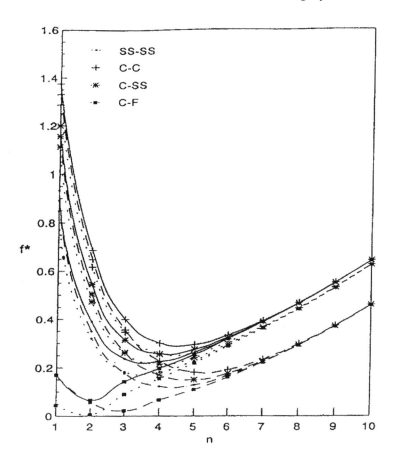

Figure 3.19 Variation of the frequency parameter $f^* = f L_s \sqrt{\rho h / A_{11}}$
with the circumferential wave number n ($m = 1$, $h/R = 0.002$,
$L/R = 6$, $\Omega = 3$ rps, solid line (—) backward wave, and dashed line
(– – –) forward wave).

rotating velocity Ω on the vibrational displacements u, v, and w. The results are presented
in Figs. 3.27–3.29.

Figure 3.18 shows the variation of the frequency parameter $f^* = f L_s \sqrt{\rho h / A_{11}}$ with
the circumferential wave number n for stationary cylindrical shells with the simply
supported–simply supported (SS–SS), clamped–clamped (C–C), clamped–simply
supported (C–SS) and clamped–free (C–F) boundary conditions. The frequencies first
decrease and then increase with circumferential wave number n. Also, at large
circumferential wave number, the frequencies converge. The convergence of the
frequencies indicates that the influence of boundary conditions is insignificant at large

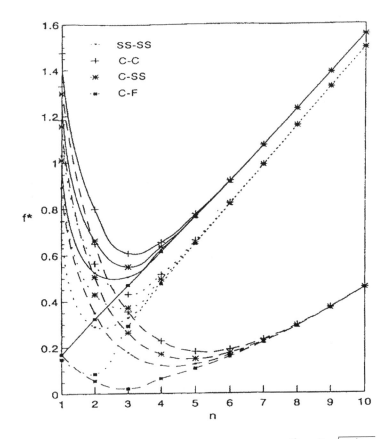

Figure 3.20 Variation of the frequency parameter $f^* = fL_s\sqrt{\rho h/A_{11}}$
with the circumferential wave number n ($m = 1$, $h/R = 0.002$,
$L/R = 6$, $\Omega = 10$ rps, solid line (—) backward wave, and dashed
line (– – –) forward wave).

circumferential wave number. For low circumferential wave numbers, C–C condition results in the highest frequencies, followed by C–SS, SS–SS and C–F.

Figures 3.19 and 3.20 show the variation of the frequency parameter f^* with the circumferential wave number n for a rotating cylindrical shell at the rotating angular velocities $\Omega = 3$ and 10 rps, respectively. Figure 3.18 has been superimposed on the two figures to illustrate the effects of rotating velocity. It is observed from the two figures that, for all circumferential wave numbers, the cylindrical shell with higher rotating velocity has the higher frequencies. For a rotating cylindrical shell, the trend of the variation of the frequencies with the circumferential wave number for the backward and forward traveling modes for the four boundary conditions are similar to that for a non-rotating cylindrical shell. The frequencies for the backward traveling

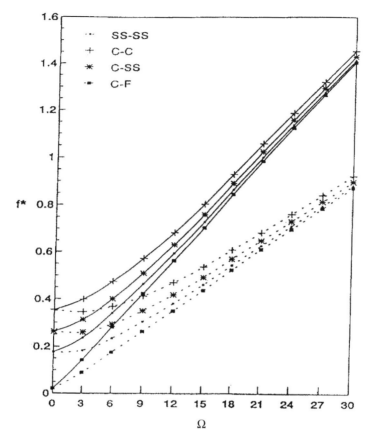

Figure 3.21 Variation of the frequency parameter $f^* = fL_s\sqrt{\rho h/A_{11}}$ with the rotating velocity Ω ($m = 1$, $n = 3$, $h/R = 0.002$, $L/R = 6$, solid line (——) backward wave, and dashed line (— — —) forward wave).

modes for the four boundary conditions are always higher than those for a non-rotating cylindrical shell. However, for the forward traveling modes, the frequencies are generally lower than those of a non-rotating cylindrical shell for small circumferential wave numbers and higher for large circumferential wave numbers. The frequencies for the backward and forward traveling modes converge faster for low circumferential wave numbers as compared with the frequencies of a non-rotating cylindrical shell. This is most significant at high rotating velocity. The C–C rotating cylindrical shell has the relative highest frequencies, followed by the corresponding C–SS, SS–SS and C–F shells. Influence of the boundary conditions is noticeably more significant at small circumferential wave numbers.

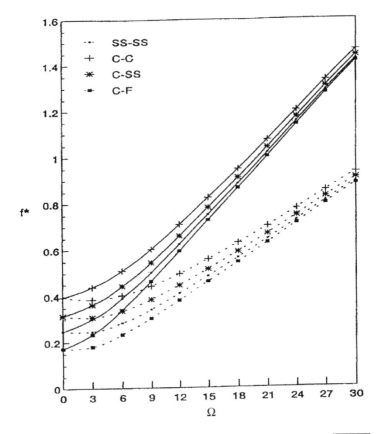

Figure 3.22 Variation of the frequency parameter $f^* = fL_s\sqrt{\rho h/A_{11}}$ with the rotating velocity Ω ($m = 1$, $n = 3$, $h/R = 0.01$, $L/R = 6$, solid line (—) backward wave, and dashed line (– – –) forward wave).

Figure 3.21 shows the variation of the shell frequency parameter $f^* = fL_s\sqrt{\rho h/A_{11}}$ with the rotating velocity Ω for the four boundary conditions SS–SS, C–C, C–SS and C–F. It is observed that the frequencies for both the backward and forward traveling waves increase monotonically with the rotating velocity, with the backward traveling waves increasing at a faster rate than the forward traveling waves. For all the rotating velocities, the C–C shell has the highest frequencies, followed by the C–SS, SS–SS and C–F shells. The influences of the boundary conditions are relatively more significant at lower rotating velocities. At very high rotating velocities, all frequencies will converge.

Figures 3.22 and 3.23 show the variation of the frequency parameter f^* with the rotating velocity Ω for thickness ratios $h/R = 0.01$ and 0.05, respectively. Comparing the frequencies in the two figures, it can be seen that for the larger h/R ratio, the frequencies

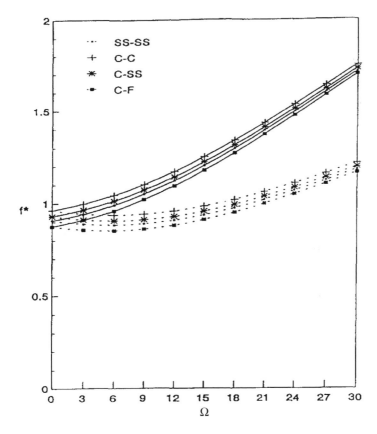

Figure 3.23 Variation of the frequency parameter $f^* = fL_s\sqrt{\rho h/A_{11}}$ with the rotating velocity Ω ($m = 1$, $n = 3$, $h/R = 0.05$, $L/R = 6$, solid line (—) backward wave, and dashed line (– – –) forward wave).

of both the backward and forward traveling waves are generally higher than those for the smaller h/R ratio, at all the rotating velocities. The increase in the frequencies with the h/R ratio could be due to an increase in the stiffness as a result of the increase in the shell thickness. The influence of the boundary conditions is relatively more significant at smaller h/R ratio.

Figure 3.24 shows the variation of the frequency parameter f^* with the rotating velocity Ω for the length ratio $L/R = 14$. Comparing this figure with Fig. 3.21 ($L/R = 6$), it can be noted that, for a longer cylindrical shell, the influence of the boundary conditions is relatively insignificant, especially at higher rotating velocities. This characteristic is again illustrated in Fig. 3.25, in which the variation of the frequency parameter f^* with

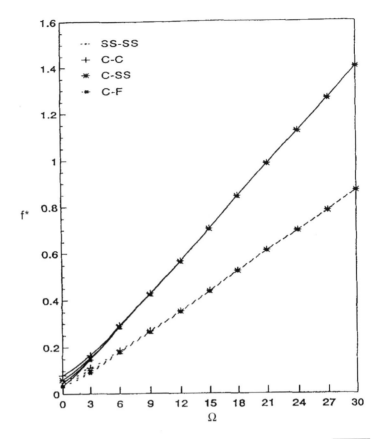

Figure 3.24 Variation of the frequency parameter $f^* = fL_s\sqrt{\rho h/A_{11}}$
with rotating velocity Ω ($m = 1$, $n = 3$, $h/R = 0.002$,
$L/R = 14$, solid line (—) backward wave, and dashed line
(– – –) forward wave).

the length ratio L/R is plotted. From the figure, noticeable influence of boundary conditions is observed for $L/R < 14$. For longer rotating cylindrical shells, the influence of boundary conditions becomes negligible. The influence of boundary conditions is most significant at the lowest L/R ratios.

Figure 3.26 shows the variation of the frequency parameter f^* with the thickness ratio h/R. It is clearly seen that the frequencies increase with the h/R ratio with the influence of boundary conditions relatively more significant at small h/R ratios. Shells with C–C boundary conditions have the highest frequencies, followed by those with C–SS, SS–SS and C–F boundary conditions.

Figures 3.27–3.29 show the relative vibrational displacements u, v, and w of a rotating cylindrical shell. For the displacement u, only the C–SS and SS–SS boundary

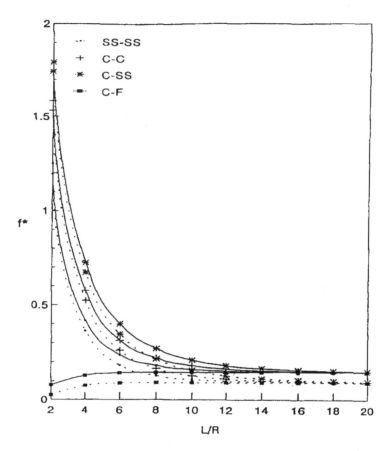

Figure 3.25 Variation of the frequency parameter $f^* = fL_s\sqrt{\rho h/A_{11}}$ with the L/R ratio ($m = 1$, $n = 3$, $h/R = 0.01$, $\Omega = 3$ rps, solid line (—) backward wave, and dashed line (– – –) forward wave).

condition cases show discrepancies in the displacements for the backward and forward traveling waves, while no noticeable discrepancies in the displacements are observed for the C–C and C–F boundary condition cases. For the displacement v, the discrepancies in the displacements for the backward and forward traveling waves are observed for all the four boundary conditions, with the C–C boundary condition case displaying the largest relative discrepancy, and this is followed by the C–F, SS–SS and C–SS boundary condition cases. For the displacement w, the discrepancies in the displacements for the backward and forward traveling waves are observable only for the C–SS boundary condition case. According to the above observations, it appears that the influence of the rotating velocity is most significant in the v-direction, as compared with the u- and w-directions.

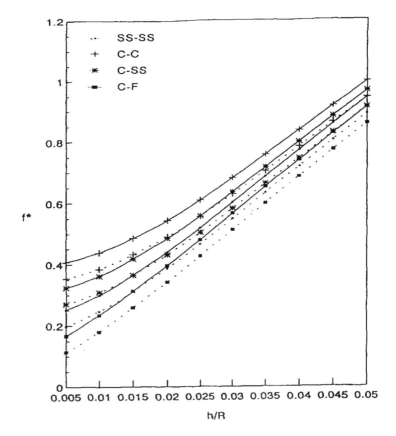

Figure 3.26 Variation of the frequency parameter $f^* = fL_s\sqrt{\rho h/A_{11}}$ with the h/R ratio ($m = 1, n = 3, L/R = 6, \Omega = 3$ rps, solid line (—) backward wave, and dashed line (– – –) forward wave).

Based on the above discussions, it can be concluded that boundary conditions have a significant influence on the frequencies of a rotating cylindrical shell. The influence is more significant for small circumferential wave numbers and/or low rotating velocities, and also for lower length and thickness ratios ($L/R, h/R$).

g)   *Discussion on modal wave numbers.*

In general, the vibrational modes of a rotating cylindrical shell consist of the longitudinal modes described by the longitudinal wave number m, and the circumferential modes described by the circumferential wave number n. Both these wave numbers represent the numbers of cycles per unit distance for a wave shape or

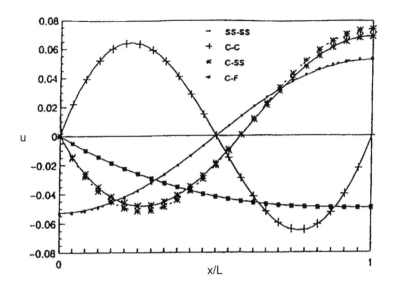

Figure 3.27 Relative displacements of the rotating laminated composite orthotropic cylindrical shell in the u-direction ($m = 1$, $n = 3$, $h/R = 0.002$, $L/R = 6$, solid line (—) backward wave, and dashed line (− − −) forward wave).

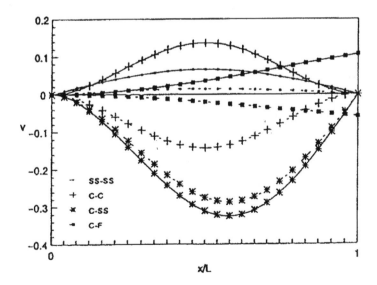

Figure 3.28 Relative displacements of the rotating laminated composite orthotropic cylindrical shell in the v-direction ($m = 1$, $n = 3$, $h/R = 0.002$, $L/R = 6$, solid line (—) backward wave, and dashed line (− − −) forward wave).

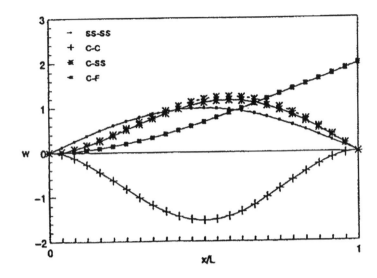

Figure 3.29 Relative displacements of the rotating laminated
composite orthotropic cylindrical shell in the w-direction ($m = 1$,
$n = 3$, $h/R = 0.002$, $L/R = 6$, solid line (—) backward wave, and
dashed line (– – –) forward wave).

waveform in the longitudinal and circumferential directions, respectively, for free
vibration of the shell of revolution. If the cylindrical shell is stationary, the vibrational
modes (m, n) present the standing waves whose vibrational amplitude varies through the
cylindrical shell but the wave shape or waveform does not propagate. When the
cylindrical shell starts to rotate, the standing wave motion is transformed to a forward or
backward traveling wave motion, depending on the rotating direction, where the constant
waveform moves through the rotating cylindrical shell. In this section, discussions are
made to better understand the characteristics of the longitudinal and circumferential
modal wave numbers (m, n) of the rotating cylindrical shell.

Figure 3.30 shows the variation of the natural frequencies f (Hz) with the rotating
velocity Ω (rps) for the vibrational modes (m, n) = (1, 2), (1, 4) and (2, 4). The present
rotating cylindrical shell with simply supported boundary condition at both edges has
three isotropic layers of construction, where the thickness of middle layer ($E =
2.0685 \times 10^{11}$ N/m^2, $\mu = 0.3$ and $\rho = 8053$ kg/m^3) is three times thicker than the two
surface layers ($E = 4.8265 \times 10^9$ N/m^2, $\mu = 0.3$ and $\rho = 1314$ kg/m^3). For the present
modes considered, the natural frequencies of the two waves increase monotonically with
the rotating velocity except for the forward waves of the mode (1, 2) which decrease
initially and then increase with the increase in the rotating velocity. All the natural
frequencies are found to increase uniformly as the rotating velocity increases.

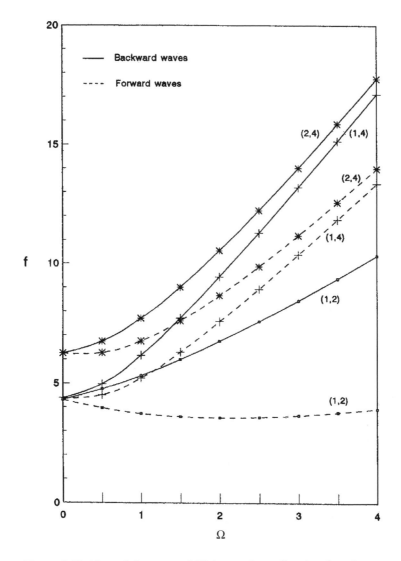

Figure 3.30 Natural frequency f (Hz) at various vibrational modes as the function of rotating velocity Ω (rps) for a simply supported rotating multi-layered cylindrical shell ($h/R = 0.002$, $L/R = 20$).

Figure 3.31 shows the variation of the non-dimensional frequency parameter $\omega^* = \omega L_s \sqrt{\rho h / A_{11}}$ (ω (rad/s) is the natural circular frequency, $L_s = 2R\pi$, R is the radius, and ρ and A_{11} are the density and extensional stiffness term) for various vibrational modes (m, n), with the rotating velocity Ω, for a cylindrical shell with the simply supported boundary condition at both edges. The cylindrical shell studied here is a thin laminated composite shell consisting of three orthotropic equithickness layers with a stacking

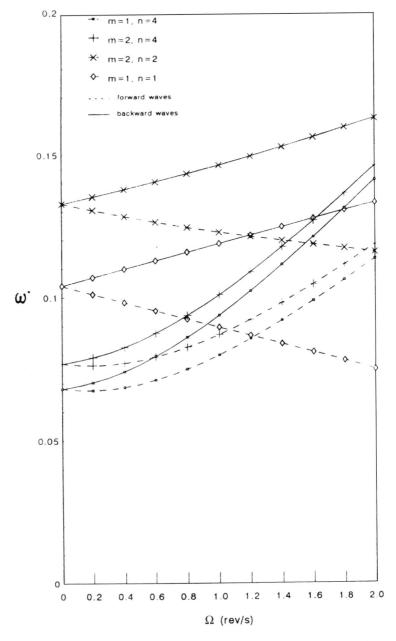

Figure 3.31 Variation of the non-dimensional frequency parameter $\omega^* = \omega L_s \sqrt{\rho h / A_{11}}$ with the rotating velocity Ω (rps) at different vibrational modes for a rotating laminated composite orthotropic cylindrical shell with simply supported boundary condition at both edges ($h/R = 0.002$, $L/R = 20$).

sequence $[90°/0°/90°]$, and the relevant material properties of each layer are $E_{22} = 7.6$ GN/m^2, $E_{11}/E_{22} = 2.5$, $G_{12} = 4.1$ GN/m^2, $\mu_{12} = 0.26$, $\rho = 1643$ kg/m^3. It is clearly seen from the figure that mode $(2,4)$ has generally higher frequency parameters for the backward and forward waves than mode $(1,4)$. The same characteristic is also observed for the mode $(2,2)$ and $(1,1)$. At high rotating velocities, the frequency parameters corresponding to the backward waves for all the four vibrational modes considered are observed to increase quasi-linearly with the rotating velocity.

Appendix A

The differential operator matrix $\bar{\mathbf{L}} = [\bar{L}_{ij}]$ (i, $j = 1$, 2, 3) in Eq. (3.31), corresponding to Love's shell theory, can be written as follows:

$$\bar{L}_{11} = A_{11}\frac{\partial^2}{\partial x^2} + \left(\frac{A_{66}}{R^2} + \rho h\Omega^2\right)\frac{\partial^2}{\partial\theta^2} - \rho h\frac{\partial^2}{\partial t^2} \tag{A.1}$$

$$\bar{L}_{12} = \left(\frac{A_{12} + A_{66}}{R} + \frac{B_{12} + 2B_{66}}{R^2}\right)\frac{\partial^2}{\partial x\,\partial\theta} \tag{A.2}$$

$$\bar{L}_{13} = -B_{11}\frac{\partial^3}{\partial x^3} + \left(\frac{A_{12}}{R} - \rho h\Omega^2 R\right)\frac{\partial}{\partial x} - \frac{B_{12} + 2B_{66}}{R^2}\frac{\partial^3}{\partial x\,\partial\theta^2} \tag{A.3}$$

$$\bar{L}_{21} = \left(\frac{A_{12} + A_{66}}{R} + \frac{B_{12} + B_{66}}{R^2} + \rho h\Omega^2 R\right)\frac{\partial^2}{\partial x\,\partial\theta} \tag{A.4}$$

$$\bar{L}_{22} = \left(A_{66} + \frac{3B_{66}}{R} + \frac{2D_{66}}{R^2}\right)\frac{\partial^2}{\partial x^2} + \left(\frac{A_{22}}{R^2} + \frac{2B_{22}}{R^3} + \frac{D_{22}}{R^4}\right)\frac{\partial^2}{\partial\theta^2}$$

$$+ \rho h\left(\Omega^2 - \frac{\partial^2}{\partial t^2}\right) \tag{A.5}$$

$$\bar{L}_{23} = -\left(\frac{B_{12} + 2B_{66}}{R} + \frac{D_{12} + 2D_{66}}{R^2}\right)\frac{\partial^3}{\partial x^2\,\partial\theta} + \left(\frac{A_{22}}{R^2} + \frac{B_{22}}{R^3}\right)\frac{\partial}{\partial\theta}$$

$$-\left(\frac{B_{22}}{R^3} + \frac{D_{22}}{R^4}\right)\frac{\partial^3}{\partial\theta^3} - 2\rho h\Omega\frac{\partial}{\partial t} \tag{A.6}$$

$$\bar{L}_{31} = B_{11}\frac{\partial^3}{\partial x^3} - \frac{A_{12}}{R}\frac{\partial}{\partial x} + \frac{B_{12} + 2B_{66}}{R^2}\frac{\partial^3}{\partial x\,\partial\theta^2} \tag{A.7}$$

$$\bar{L}_{32} = \left(\frac{B_{12} + 2B_{66}}{R} + \frac{D_{12} + 4D_{66}}{R^2}\right)\frac{\partial^3}{\partial x^2 \partial \theta} - \left(\frac{A_{22}}{R^2} + \frac{B_{22}}{R^3} + \rho h \Omega^2\right)\frac{\partial}{\partial \theta}$$

$$+ \left(\frac{B_{22}}{R^3} + \frac{D_{22}}{R^4}\right)\frac{\partial^3}{\partial \theta^3} + 2\rho h \Omega \frac{\partial}{\partial t} \tag{A.8}$$

$$\bar{L}_{33} = -D_{11}\frac{\partial^4}{\partial x^4} + \frac{2B_{12}}{R}\frac{\partial^2}{\partial x^2} - \frac{2D_{12} + 4D_{66}}{R^2}\frac{\partial^4}{\partial x^2 \partial \theta^2} + \left(\frac{2B_{22}}{R^3} + \rho h \Omega^2\right)\frac{\partial^2}{\partial \theta^2}$$

$$- \frac{D_{22}}{R^4}\frac{\partial^4}{\partial \theta^4} - \left(\frac{A_{22}}{R^2} - \rho h \Omega^2 + \rho h \frac{\partial^2}{\partial t^2}\right) \tag{A.9}$$

Chapter 4
Free Vibration of Thin Rotating Conical Shells

4.1   Introduction.

An extensive search of the literature will reveal very few studies on rotating conical shells. Many studies have, however, been carried out on the dynamic analysis of stationary conical shell.

Employing the Rayleigh–Ritz method, Bacon & Bert [1967] carried out the free vibration analysis of both isotropic and orthotropic non-rotating conical shells. Sankaranarayanan et al. [1987] studied the axisymmetric free vibration of laminated non-rotating conical shells with linear thickness variation in the meridional direction. The effect of transverse shear deformation has also been included in further studies by Kayran & Vinson [1990] using a combination of modal iteration and transfer matrix approach, and by Khdeir & Reddy [1990] who used the third-order shear deformation shell theory of Reddy [1984]. Through a numerical integration method, Irie et al. [1984] tabulated the natural frequencies for truncated circular conical shells with different boundary conditions. Srinivasan & Krishnan [1989] used a similar integral equation method in the space domain to carry out a dynamic response analysis of stiffened conical shell panels. Based on different versions of the finite element method (FEM), Ross [1975], Chang et al. [1983], Sundarasivarao & Ganesan [1991], Singh et al. [1991] and Thambiratnam & Zhuge [1993] studied the vibration of the various stationary conical shells. Sivadas & Ganesan [1991, 1992] analyzed the free vibration of thin as well as thick laminated composite conical shells of varying thickness, where Love's [1927] first approximation was used for thin conical shells with linear symmetrically varying thickness. Also, a higher order shell theory, which includes the thickness normal strain and two transverse strains, was used for the vibration analysis of thick clamped conical shells. In addition to the above works, a power series solution procedure was presented by Tong [1993a,b] for the free vibration analysis of isotropic, orthotropic and laminated composite conical shells.

For the dynamic analysis of the rotating conical shells, the first notable work is attributed to Sivadas [1995] who studied the free vibration of a moderately thick rotating circular conical shell with material damping by the FEM. This was followed by works from the authors of this monograph, Lam and Li [Lam & Li, 1997, 1999a,b, 2000a,b; Lam et al., 2002, 2003; Li, 2000a–c; Li & Lam, 1998, 2000, 2001; Ng et al.,

1999, 2003], who conducted a comprehensive and systematic study, over several years, on the free vibration of rotating thin truncated circular conical shells using both the Galerkin-based analytical approach and the generalized differential quadrature (GDQ)-based assumed-mode method. In this chapter, the important aspects of this extensive study are summarized and the influences of rotating speed, cone angle, geometric parameter, material property, meridional wave number, circumferential wave number, boundary condition and initial stress, on the frequency characteristics, are elucidated. For the validation of the accuracy of the implemented methodologies, various comparisons are made with those available in open literature. Significant findings are highlighted and discussed in detail.

4.2   Theoretical development: rotating conical shell.

For a thin truncated circular conical shell, as shown in Fig. 4.1, rotating at a constant angular velocity Ω about its symmetrical and horizontal axis, α is the cone angle and $\alpha + \varphi = 90°$, L is the length and h is the thickness of the conical shell; a and b are the mean radii at both ends. The reference surface for the deformations of the rotating conical shell is taken to be the middle surface on which an orthogonal coordinate system (x, θ, z) is fixed. $r = r(x) = a + x \sin \alpha$ is the mean radius at any coordinate point (x, θ, z). Deformations of the rotating conical shell are defined by u, v, and w in the meridional x, circumferential θ, and normal z-directions, respectively.

From the definitions of the Lamé parameters, A_1 and A_2 can be made to be dimensional or non-dimensional, depending on the dimensions of the curvilinear coordinates α_1 and α_2. For the present rotating conical shell, if the transformations of

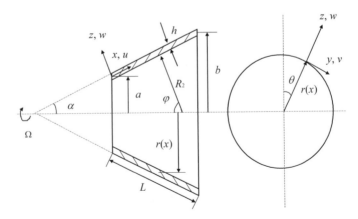

Figure 4.1 Geometry and coordinate system of a rotating conical shell.

Table 4.1

Comparison of frequency parameter $f^* = \omega R\sqrt{(1 - \mu^2)\rho/E}$ for non-rotating single-layer isotropic cylindrical shell with simply supported–simply supported boundary condition ($m = 1$, $\mu = 0.3$, $R = a = b$, $R/L = 0.05$, $h/R = 0.05$) using the two distributions of discrete grid points.

n	Markus [1988] (3D elasticity solution)	NGP	Present (the GDQ method)	
			Cosine distribution	Equidistant distribution
2	0.039233	7	0.032637	0.036841
		8	0.039464	0.038998
		10	0.039811	0.039968
		16	0.039829	0.040079
		21	0.039819	0.040007
3	0.109477	7	0.110168	0.109248
		8	0.107265	0.109523
		10	0.109866	0.109594
		16	0.109898	0.109946
		21	0.109897	0.109946
4	0.209008	7	0.209455	0.210094
		8	0.209578	0.210233
		10	0.210247	0.210231
		16	0.210310	0.210307
		21	0.210310	0.210342

NGP refers to the total number of grid points in the meridional direction

Table 4.2

Comparison of frequency parameter $f^* = \omega R\sqrt{(1 - \mu^2)\rho/E}$ for non-rotating single-layer isotropic cylindrical shell with clamped–clamped boundary condition ($m = 1$, $\mu = 0.3$, $R = a = b$).

Case	Dym [1973][a]	Chung [1981][b]		Present (the GDQ method)		
		20 terms	500 terms	NGP = 6	NGP = 10	NGP = 16
$L/R = 10$, $R/h = 500$, $n = 4$	0.01508	0.01515	0.01508	0.01672	0.01532	0.01516
$L/R = 10$, $R/h = 20$, $n = 2$	0.05787	0.05795	0.05784	0.06226	0.05856	0.05786

NGP refers to the total number of grid points in the meridional direction

[a] Exact solution

[b] Series solution

Table 4.3

Comparison of theoretically computed and experimentally measured frequencies (cps) for a non-rotating isotropic circular cylindrical shell with clamped–clamped boundary condition, and subjected to an initial internal uniform pressure ($R = a = b$, $L = l$).

Material/geometry	p_0	n	Experimental data	GDQ method	Relative discrepancy (%)
17-7 PH,	0	5	160	172.11	7.0
stainless steel,	0	6	189	193.74	2.4
$R = 6.01$ in., $L = 36$ in.,	0	7	239	245.30	2.5
$h = 0.0185$ in.	0	8	318	314.37	1.2
	0	9	399	396.14	0.75
	4	3	262	290.71	9.8
	2	4	193	174.42	9.8
	6	5	256	268.85	4.7
	8	6	335	364.49	8.0
2024 aluminum,	3	4	322	305.57	5.2
$R = 6.01$ in., $L = 36$ in.,	3	5	356	385.29	7.6
$h = 0.0060$ in.	3	6	443	407.39	8.1
301 stainless steel,	4	2	432	476.29	9.2
$R = 6.01$ in.,	6	2	429	464.68	7.6
$L = 38.20$ in.,	8	2	434	452.78	4.1
$h = 0.0037$ in.	10	2	432	440.55	1.9

p_0 is the initial internal uniform pressure and its unit is psi. The present experimental data are quoted from the report by Miserentino & Vosteen [1965], "Vibration Tests of Pressurized Thin-Walled Cylindrical Shells", NASA TN D 3066. The total number of discrete mesh points is taken to be 10 here

the curvilinear coordinate systems are taken as

$$\alpha_1 = x \quad \text{and} \quad \alpha_2 = \theta \tag{4.1}$$

the Lamé parameters will be

$$A_1 = 1 \quad \text{and} \quad A_2 = r(x) = a + x \sin \alpha \tag{4.2}$$

From the definition of the radii R_1 and R_2 described above, it is obvious that the radius R_1 is infinity, and the radius R_2 is also obtained as

$$R_1 = \infty \quad \text{and} \quad R_2 = r(x)/\cos \alpha \tag{4.3}$$

Based on these descriptions, substituting Eqs. (4.1)–(4.3) into Eqs. (2.62)–(2.67) and (2.68)–(2.73), two sets of resulting equations are obtained, respectively, for the basic and additional states of the rotating conical shell.

Table 4.4

Comparison of frequency parameter $f_b^* = \omega R\sqrt{(1 - \mu^2)\rho/E}$ of backward wave for an infinitely long rotating single-layer isotropic cylindrical shell ($m = 1$, $\mu = 0.3$, $R = a = b$, $h/R = 0.002$).

Ω (rps)	n	Chen et al. [1993][a]	Present (the GDQ method)		
			NGP = 6	NGP = 10	NGP = 16
0.05	2	0.0016802	0.0016805	0.0016903	0.0016903
	3	0.0044779	0.0044921	0.0044763	0.0044760
	4	0.0084766	0.0084941	0.0084892	0.0084554
	5	0.0136483	0.0136672	0.0136653	0.0136516
	6	0.0199836	0.0200033	0.0200023	0.0199958
	7	0.0274785	0.0274985	0.0274980	0.0274945
	8	0.0361310	0.0361513	0.0361510	0.0361489
	9	0.0459401	0.0459606	0.0459604	0.0459592
	10	0.0569052	0.0569258	0.0569257	0.0569249
0.1	2	0.0018232	0.0018529	0.0018625	0.0018625
	3	0.0045788	0.0046403	0.0046246	0.0045144
	4	0.0085541	0.0086256	0.0086208	0.0085872
	5	0.0137109	0.0137871	0.0137852	0.0137716
	6	0.0200361	0.0201148	0.0201139	0.0201074
	7	0.0275236	0.0276038	0.0276033	0.0275998
	8	0.0361705	0.0362518	0.0362515	0.0362495
	9	0.0459752	0.0460572	0.0460570	0.0460558
	10	0.0569368	0.0570193	0.0570192	0.0570184

NGP refers to the total number of grid points in the meridional direction. Subscript "b" denotes backward wave
[a] Equation from Chen et al. [1993]:

$$\omega_b^* = \frac{2n}{n^2 + 1}\Omega + \sqrt{\frac{n^2(n^2 - 1)^2}{n^2 + 1}\frac{Eh^2}{12(1 - \mu^2)\rho R^2} + \frac{n^4 + 3}{(n^2 + 1)^2}\Omega^2}$$

For the basic state of the rotating conical shell, the resulting set of equations is simplified and expressed as

$$\frac{\partial(r(x)N_{Bx})}{\partial x} + \frac{\partial N_{B\theta x}}{\partial \theta} - N_{B\theta}\sin\alpha + \rho h\Omega^2 r^2(x)\sin\alpha = 0 \tag{4.4}$$

$$\frac{\partial(r(x)N_{Bx\theta})}{\partial x} + \frac{\partial N_{B\theta}}{\partial \theta} + N_{B\theta x}\sin\alpha + Q_{B\theta}\cos\alpha = 0 \tag{4.5}$$

$$\frac{\partial(r(x)Q_{Bx})}{\partial x} + \frac{\partial Q_{B\theta}}{\partial \theta} - N_{B\theta}\cos\alpha + \rho h\Omega^2 r^2(x)\cos\alpha = 0 \tag{4.6}$$

Table 4.5

Comparison of frequency parameter $f_f^* = \omega R\sqrt{(1-\mu^2)\rho/E}$ of forward wave for an infinitely long rotating single-layer isotropic cylindrical shell ($m = 1$, $\mu = 0.3$, $R = a = b$, $h/R = 0.002$).

Ω (rps)	n	Chen et al. [1993][a]	Present (the GDQ method)		
			NGP = 6	NGP = 10	NGP = 16
0.05	2	0.0014301	0.0014303	0.0014401	0.0014401
	3	0.0042903	0.0043045	0.0042886	0.0043062
	4	0.0083294	0.0083469	0.0083421	0.0083082
	5	0.0135280	0.0135469	0.0135449	0.0135313
	6	0.0198822	0.0199018	0.0199008	0.0198944
	7	0.0273909	0.0274109	0.0274105	0.0274069
	8	0.0360540	0.0360743	0.0360740	0.0360720
	9	0.0458715	0.0458919	0.0458918	0.0458905
	10	0.0568433	0.0568638	0.0568637	0.0568629
0.1	2	0.0013228	0.0013526	0.0013621	0.0013621
	3	0.0042035	0.0042650	0.0042493	0.0041391
	4	0.0082597	0.0083313	0.0083264	0.0082928
	5	0.0134704	0.0135466	0.0135446	0.0135310
	6	0.0198332	0.0199119	0.0199110	0.0199045
	7	0.0273484	0.0274288	0.0274282	0.0274247
	8	0.0360165	0.0360978	0.0360975	0.0360955
	9	0.0458379	0.0459199	0.0459197	0.0459185
	10	0.0568129	0.0568954	0.0568953	0.0568945

NGP refers to the total number of grid points in the meridional direction. Subscript "f" denotes forward wave
[a] Equation from Chen *et al.* [1993]:

$$\omega_f^* = \frac{2n}{n^2+1}\Omega - \sqrt{\frac{n^2(n^2-1)^2}{n^2+1}\frac{Eh^2}{12(1-\mu^2)\rho R^2} + \frac{n^4+3}{(n^2+1)^2}\Omega^2}$$

$$\frac{\partial(r(x)M_{Bx})}{\partial x} + \frac{\partial M_{B\theta x}}{\partial \theta} - M_{B\theta}\sin\alpha - Q_{Bx}r(x) = 0 \tag{4.7}$$

$$\frac{\partial(r(x)M_{Bx\theta})}{\partial x} + \frac{\partial M_{B\theta}}{\partial \theta} + M_{B\theta x}\sin\alpha - Q_{B\theta}r(x) = 0 \tag{4.8}$$

$$M_{B\theta x}\cos\alpha - r(x)(N_{Bx\theta} - N_{B\theta x}) = 0 \tag{4.9}$$

It is noted that there is no time component in the set of resulting equations (4.4)–(4.9). Therefore, the basic state equations of the rotating conical shell can be

Table 4.6

Comparison of frequency parameters f_b^*, $f_f^* = \omega R\sqrt{(1 - \mu^2)\rho/E}$ of both backward and forward waves for an infinitely long rotating cylindrical shell ($m = 1$, $\mu = 0.3$, $h/R = 0.002$, $R = a = b$).

Ω (rps)	n	Chen et al. [1993][a]		Lam et al. [1995a][b]		Present (GDQ)	
		f_b^*	f_f^*	f_b^*	f_f^*	f_b^*	f_f^*
0.05	2	0.00168	0.00143	0.00169	0.00144	0.00169	0.00144
	3	0.00448	0.00429	0.00449	0.00431	0.00448	0.00431
	4	0.00848	0.00833	0.00850	0.00835	0.00846	0.00831
	5	0.01365	0.01353	0.01370	0.01355	0.01365	0.01353
0.1	2	0.00182	0.00132	0.00186	0.00136	0.00186	0.00136
	3	0.00458	0.00420	0.00464	0.00427	0.00451	0.00414
	4	0.00855	0.00826	0.00863	0.00833	0.00859	0.00829
	5	0.01371	0.01347	0.01379	0.01355	0.01377	0.01353

Subscripts "b" and "f" denote the backward and forward waves, respectively
[a] Equation from Chen et al. [1993]:

$$\omega_{b,f}^* = \frac{2n}{n^2 + 1}\Omega \pm \sqrt{\frac{n^2(n^2 - 1)^2}{n^2 + 1}\frac{Eh^2}{12(1 - \mu^2)\rho R^2} + \frac{n^4 + 3}{(n^2 + 1)^2}\Omega^2}$$

[b] The results were obtained directly from the governing equations of motion by using the trigonometric functions as the trial functions of the displacement field

taken as the initial state equations for the free vibration of the shell. For ease of analysis, free boundary conditions are imposed on both edges of the shell, $x = 0$ and L, for the basic state equations. In this way, the solutions of the initial state equations, namely those of the basic state equations, for the free vibration of the rotating shell are obtained by

$$N_{B\theta} = \rho h\Omega^2 r^2(x) = \rho h\Omega^2(a + x\sin\alpha)^2, \qquad N_{Bx} = N_{Bx\theta} = N_{B\theta x} = 0 \qquad (4.10)$$

$$M_{Bx} = M_{B\theta} = M_{Bx\theta} = M_{B\theta x} = 0, \qquad Q_{Bx} = Q_{B\theta} = 0 \qquad (4.11)$$

Here, $N_{B\theta}$ is the only nonzero solution and represents the hoop tension due to the rotating velocity Ω. It will now be denoted by N_θ^0 and is defined as the initial hoop

Table 4.7

Comparison of frequency parameter $f = \omega b \sqrt{(1 - \mu^2)\rho/E}$ for non-rotating single-layer isotropic conical shells with Cs–Cl and Ss–Sl boundary conditions ($m = 1$, $\mu = 0.3$, $\alpha = 30°$, $h/b = 0.01$).

			Present (the GDQ method)		
Case	n	Irie *et al.* [1984][a]	NGP = 6	NGP = 10	NGP = 16
Cs–Cl	1	0.7526	0.7822	0.7506	0.7489
$L \sin \alpha/b = 0.75$	2	0.4662	0.5092	0.4669	0.4618
	3	0.3164	0.3508	0.3165	0.3110
	4	0.2619	0.2778	0.2593	0.2564
	5	0.2667	0.2750	0.2626	0.2606
	6	0.2953	0.3036	0.2899	0.2873
Ss–Sl	2	0.6189	0.6199	0.6180	0.6180
$L \sin \alpha/b = 0.5$	3	0.4157	0.4123	0.4112	0.4111
	4	0.2988	0.2940	0.2938	0.2936
	5	0.2535	0.2460	0.2483	0.2484
	6	0.2612	0.2513	0.2559	0.2561
	7	0.2996	0.2894	0.2940	0.2943

NGP refers to the total number of grid points in the meridional direction
[a] Numerical integration method

Table 4.8

Comparison of frequency parameter $f = \omega b \sqrt{(1 - \mu^2)\rho/E}$ for a non-rotating single-layer isotropic conical shell with Cs–Cl boundary condition ($m = 1$, $\mu = 0.3$, $h/b = 0.01$).

	$\alpha = 45°$, $L \sin \alpha/b = 0.5$		$\alpha = 60°$, $L \sin \alpha/b = 0.75$	
n	Irie *et al.* [1984][a]	GDQ method	Irie *et al.* [1984][a]	GDQ method
2	0.6696	0.6579	0.4656	0.4653
3	0.5430	0.5384	0.3614	0.3662
4	0.4570	0.4547	0.3155	0.3164
5	0.4095	0.4068	0.3192	0.3140
6	0.3970	0.3924	0.3545	0.3441
7	0.4151	0.4083	0.4082	0.3940

[a] Numerical integration method

Table 4.9

Comparison of frequency parameter $f = \omega b\sqrt{(1 - \mu^2)\rho/E}$ for a non-rotating single-layer isotropic conical shell with Ss–Sl boundary condition ($m = 1$, $\mu = 0.3$, $h/b = 0.01$, $L \sin \alpha/b = 0.75$).

	$\alpha = 30°$		$\alpha = 45°$		$\alpha = 60°$	
n	Irie et al. [1984][a]	GDQ	Irie et al. [1984][a]	GDQ	Irie et al. [1984][a]	GDQ
2	0.3598	0.3550	0.5022	0.4959	0.4450	0.4406
3	0.2134	0.2066	0.3086	0.2999	0.3080	0.2978
4	0.1932	0.1846	0.2495	0.2388	0.2574	0.2481
5	0.2183	0.2117	0.2618	0.2535	0.2703	0.2610
6	0.2543	0.2487	0.2972	0.2893	0.3089	0.2974
7	0.2995	0.2938	0.3441	0.3353	0.3614	0.3497
8	0.3533	0.3470	0.4014	0.3988	0.4250	0.4207

[a] Numerical integration method

tension due to the centrifugal force effect

$$N_\theta^0 = \rho h\Omega^2 r^2(x) = \rho h\Omega^2(a + x \sin \alpha)^2 \tag{4.12}$$

Further, for the additional state of the rotating conical shell, using the solutions of

Table 4.10

Comparison of frequency parameter $f = \omega b\sqrt{(1 - \mu^2)\rho/E}$ for a non-rotating single-layer isotropic conical shell with Ss–Cl boundary condition ($m = 1$, $\mu = 0.3$, $h/b = 0.01$).

		$L \sin \alpha/b = 0.5$		$L \sin \alpha/b = 0.75$	
Cone angle	n	Irie et al. [1984][a]	GDQ method	Irie et al. [1984][a]	GDQ method
$\alpha = 30°$	1	0.8760	0.8808	0.7451	0.7452
	2	0.6278	0.6241	0.4510	0.4530
	3	0.4574	0.4537	0.2990	0.2998
	4	0.3551	0.3506	0.2545	0.2520
	5	0.3073	0.3017	0.2661	0.2620
$\alpha = 45°$	5	0.3834	0.3804	0.3150	0.3147
	6	0.3747	0.3691	0.3437	0.3475
	7	0.3997	0.3924	0.3883	0.3898
	8	0.4489	0.4411	0.4454	0.4458
	9	0.5142	0.5063	0.5133	0.5195

[a] Numerical integration method

Table 4.11

Comparison of frequency parameter $f = \omega b \sqrt{(1 - \mu^2)\rho/E}$ for a non-rotating single-layer isotropic conical shell with Cs–Sl boundary condition ($m = 1$, $\mu = 0.3$, $h/b = 0.01$).

	$\alpha = 45°$, $L \sin \alpha/b = 0.5$		$\alpha = 60°$, $L \sin \alpha/b = 0.75$	
n	Irie et al. [1984][a]	GDQ method	Irie et al. [1984][a]	GDQ method
0	0.7149	0.7088	0.4741	0.4643
1	0.7095	0.7079	0.4961	0.4966
2	0.6474	0.6494	0.4512	0.4579
3	0.5203	0.5188	0.3210	0.3221
4	0.4164	0.4135	0.2675	0.2624
5	0.3598	0.3556	0.2734	0.2626

[a] Numerical integration method

the basic state equations (4.10)–(4.12), the set of resulting equations is expressed by

$$\frac{1}{r(x)}\left(\frac{\partial(r(x)N_{Ax})}{\partial x} + \frac{\partial N_{A\theta x}}{\partial \theta} - N_{A\theta}\sin\alpha + N_\theta^0 e_1 \sin\alpha\right) - \rho h \Omega^2 r(x) e_1 \sin\alpha$$
$$+ \rho h \Omega^2 r(x) \vartheta \cos\alpha + 2\rho h \Omega \frac{\partial v}{\partial t}\sin\alpha - \rho h \frac{\partial^2 u}{\partial t^2} = 0 \qquad (4.13)$$

$$\frac{1}{r(x)}\left(\frac{\partial(r(x)N_{Ax\theta})}{\partial x} + \frac{\partial N_{A\theta}}{\partial \theta} + N_{A\theta x}\sin\alpha + Q_{A\theta}\cos\alpha\right)$$
$$+ \frac{N_\theta^0}{r(x)}\left(\frac{\partial e_1}{\partial \theta} + \beta_2 \sin\alpha - \psi \cos\alpha\right) + \rho h \Omega^2 r(x)(\psi \cos\alpha + \beta_1 \sin\alpha)$$
$$- 2\rho h \Omega \left(\frac{\partial u}{\partial t}\sin\alpha + \frac{\partial w}{\partial t}\cos\alpha\right) - \rho h \frac{\partial^2 v}{\partial t^2} = 0 \qquad (4.14)$$

$$\frac{1}{r(x)}\left(\frac{\partial(r(x)Q_{Ax})}{\partial x} + \frac{\partial Q_{A\theta}}{\partial \theta} - \frac{\partial(N_\theta^0 \psi)}{\partial \theta} - N_{A\theta}\cos\alpha + N_\theta^0 e_2 \cos\alpha\right)$$
$$- \rho h \Omega^2 r(x) e_1 \cos\alpha + 2\rho h \Omega \frac{\partial w}{\partial t}\cos\alpha - \rho h \frac{\partial^2 w}{\partial t^2} = 0 \qquad (4.15)$$

$$\frac{1}{r(x)}\left(\frac{\partial(r(x)M_{Ax})}{\partial x} + \frac{\partial M_{A\theta x}}{\partial \theta} - M_{A\theta}\sin\alpha\right) - Q_{Ax} = 0 \qquad (4.16)$$

Table 4.12

Comparison of theoretically computed and experimentally measured frequencies (cps) for a non-rotating aluminum conical shell with Ss–Sl boundary condition and subjected to an initial external uniform pressure ($p_0/p_{cr} = 0.446$, $m = 1$, $\alpha = 20°$, $a = 2.144$ in., $l = L \cos \alpha = 8.00$ in., $h = 0.020$ in.).

n	Experiment data	Weingarten [1966][a]	Present (the GDQ method)				Relative discrepancy[b] (%)
			NGP = 8	NGP = 10	NGP = 16	NGP = 26	
4	1163	862	818.5924	819.2947	815.9689	814.6769	30.0
5	944	609	661.9265	665.5012	665.0466	664.4245	29.6
6	840	569	688.9673	693.1407	694.8536	694.8699	17.3
7	880	651	799.1974	801.8304	804.3383	804.6870	8.6
8	985	771	933.2413	933.7303	936.3129	936.7393	4.9
9	1130	912	1080.569	1081.672	1084.093	1084.528	4.0
10	1301	1075	1241.096	1247.918	1249.626	1250.054	3.9
11	–	1260	1417.263	1432.467	1433.009	1433.422	–
12	–	1466	1612.664	1634.258	1634.208	1634.596	–
13	1949	1692	1830.223	1852.362	1853.219	1853.572	4.9
14	2204	1937	2071.486	2086.738	2090.014	2090.324	5.2

p_0 is the initial external uniform pressure and p_{cr} is the critical pressure for buckling so that p_0/p_{cr} can be defined as a non-dimensional initial pressure. l is the length in the symmetrical axis direction and hence $l = L \cos \alpha$. NGP represents the number of total discrete mesh points in the meridional direction. The experimental data are quoted from Weingarten's [1966] work

[a] In his theoretical calculation, the meridional and circumferential displacements u and v were assumed to be dependent on the normal displacement w such that the governing equations of motion were simplified to a set of uncoupled equations in terms of w. The Galerkin method was then used

[b] The present relative discrepancy is defined as that between the experimental data and the computed numerical results using the GDQ method for NGP = 26

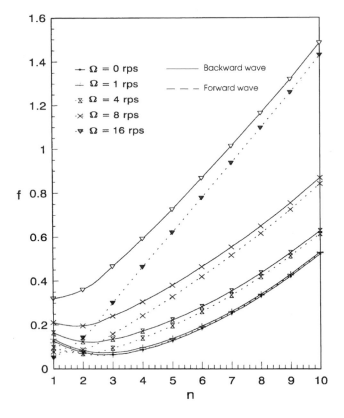

Figure 4.2 Relationship between the frequency parameter f and circumferential wave number n at various rotating speeds for cone angle $\alpha = 5°$ ($m = 1$, $\mu = 0.3$, $h/a = 0.02$, $L/a = 20$).

$$\frac{1}{r(x)}\left(\frac{\partial(r(x)M_{Ax\theta})}{\partial x} + \frac{\partial M_{A\theta}}{\partial \theta} + M_{A\theta x}\sin\alpha\right) - Q_{A\theta} = 0 \tag{4.17}$$

$$-\frac{M_{A\theta x}\cos\alpha}{r(x)} + N_{Ax\theta} - N_{A\theta x} = 0 \tag{4.18}$$

From Eqs. (4.16) and (4.17) we have

$$Q_{Ax} = \frac{1}{r(x)}\left(\frac{\partial(r(x)M_{Ax})}{\partial x} + \frac{\partial M_{A\theta x}}{\partial \theta} - M_{A\theta}\sin\alpha\right) \tag{4.19}$$

$$Q_{A\theta} = \frac{1}{r(x)}\left(\frac{\partial(r(x)M_{Ax\theta})}{\partial x} + \frac{\partial M_{A\theta}}{\partial \theta} + M_{A\theta x}\sin\alpha\right) \tag{4.20}$$

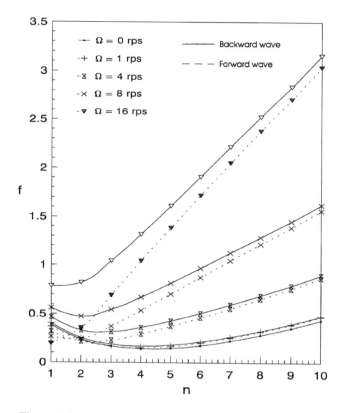

Figure 4.3 Relationship between the frequency parameter
f and circumferential wave number n at various rotating
speeds for cone angle $\alpha = 15°$ ($m = 1$, $\mu = 0.3$,
$h/a = 0.02$, $L/a = 20$).

From the general strain–displacement relationships of shells of revolution defined earlier in Eqs. (2.46)–(2.48), and considering Eqs. (4.1)–(4.3), the geometric deformation relationships of the rotating conical shell are obtained as

$$e_1 = \frac{\partial u}{\partial x}, \qquad e_2 = \frac{1}{r(x)}\frac{\partial v}{\partial \theta} + \frac{u \sin\alpha + w \cos\alpha}{r(x)} \qquad\qquad (4.21)$$

$$e_{12} = \beta_1 + \beta_2, \qquad \beta_1 = \frac{\partial v}{\partial x}, \qquad \beta_2 = \frac{1}{r(x)}\frac{\partial u}{\partial \theta} - \frac{v \sin\alpha}{r(x)} \qquad\qquad (4.22)$$

$$\vartheta = -\frac{\partial w}{\partial x}, \qquad \psi = -\frac{1}{r(x)}\frac{\partial w}{\partial \theta} + \frac{v \cos\alpha}{r(x)} \qquad\qquad (4.23)$$

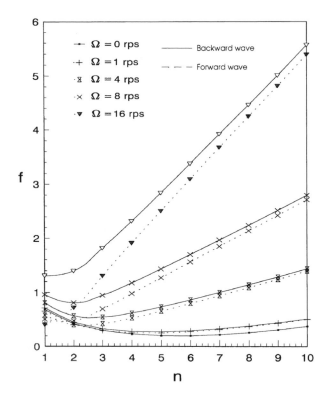

Figure 4.4 Relationship between the frequency parameter
f and circumferential wave number *n* at various rotating
speeds for cone angle $\alpha = 30°$ ($m = 1$, $\mu = 0.3$,
$h/a = 0.02$, $L/a = 20$).

Substituting Eqs. (4.19)–(4.23), with $r = r(x) = a + x \sin \alpha$, into Eqs. (4.13)–(4.15), and then simplifying, the resulting expressions in which the subscript A is omitted for ease of later discussion, the governing equations of motion in terms of forces, moments and displacements are finally derived for the free vibration of a rotating conical shell and expressed as follows:

$$\frac{\partial N_x}{\partial x} + \frac{1}{r(x)} \frac{\partial N_{x\theta}}{\partial \theta} + \frac{N_\theta^0}{r^2(x)} \left(\frac{\partial^2 u}{\partial \theta^2} - r(x) \cos \alpha \frac{\partial w}{\partial x} \right) + \frac{\sin \alpha}{r(x)} (N_x - N_\theta)$$

$$+ 2\rho h \Omega \sin \alpha \frac{\partial v}{\partial t} - \rho h \frac{\partial^2 u}{\partial t^2} = 0 \qquad (4.24)$$

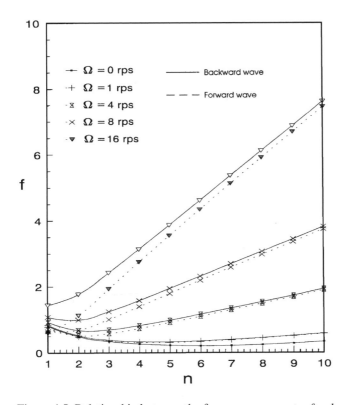

Figure 4.5 Relationship between the frequency parameter f and circumferential wave number n at various rotating speeds for cone angle $\alpha = 45°$ ($m = 1$, $\mu = 0.3$, $h/a = 0.02$, $L/a = 20$).

$$\frac{\partial N_{x\theta}}{\partial x} + \frac{1}{r(x)}\frac{\partial N_\theta}{\partial \theta} + \frac{\cos \alpha}{r(x)}\frac{\partial M_{x\theta}}{\partial x} + \frac{\cos \alpha}{r^2(x)}\frac{\partial M_\theta}{\partial \theta}$$

$$+ \frac{N_\theta^0}{r^2(x)}\left(r(x)\frac{\partial^2 u}{\partial x\,\partial \theta} + \sin \alpha\frac{\partial u}{\partial \theta} + r(x)\sin \alpha\frac{\partial v}{\partial x} - v\sin^2\alpha\right)$$

$$+ 2\frac{\sin \alpha}{r(x)}N_{x\theta} - 2\rho h\Omega\left(\sin \alpha\frac{\partial u}{\partial t} + \cos \alpha\frac{\partial w}{\partial t}\right) - \rho h\frac{\partial^2 v}{\partial t^2} = 0 \qquad (4.25)$$

$$\frac{\partial^2 M_x}{\partial x^2} + \frac{2}{r(x)}\frac{\partial^2 M_{x\theta}}{\partial \theta\,\partial x} + \frac{1}{r^2(x)}\frac{\partial^2 M_\theta}{\partial \theta^2} + \frac{2\sin \alpha}{r(x)}\frac{\partial M_x}{\partial x} - \frac{\sin \alpha}{r(x)}\frac{\partial M_\theta}{\partial x}$$

$$+ \frac{N_\theta^0}{r^2(x)}\left(\frac{\partial^2 w}{\partial \theta^2} - r(x)\cos \alpha\frac{\partial u}{\partial x}\right) + \frac{N_\theta^0}{r^2(x)}(w\cos^2\alpha + u\sin \alpha\cos \alpha)$$

$$- \frac{\cos \alpha}{r(x)}N_\theta + 2\rho h\Omega\cos \alpha\frac{\partial v}{\partial t} - \rho h\frac{\partial^2 w}{\partial t^2} = 0 \qquad (4.26)$$

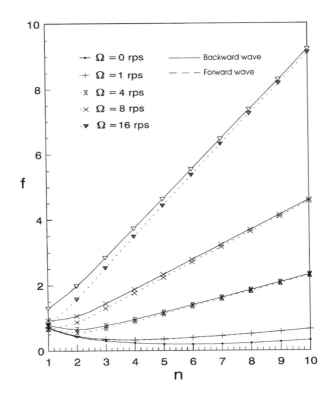

Figure 4.6 Relationship between the frequency parameter
f and circumferential wave number n at various rotating
speeds for cone angle $\alpha = 60°$ ($m = 1$, $\mu = 0.3$,
$h/a = 0.02$, $L/a = 20$).

where the three terms, $\partial^2 u/\partial t^2$, $\partial^2 v/\partial t^2$ and $\partial^2 w/\partial t^2$, on the left-hand side of
Eqs. (4.24)–(4.26) are the relative accelerations. The four terms, $2\Omega \sin \alpha\, \partial v/\partial t$,
$2\Omega \sin \alpha\, \partial u/\partial t$, $2\Omega \cos \alpha\, \partial w/\partial t$ and $2\Omega \cos \alpha\, \partial v/\partial t$, are the Coriolis accelerations.
The implicit terms, $u\Omega^2$, $v\Omega^2$, and $w\Omega^2$ are the centrifugal accelerations.

If an initial stress is applied to the rotating isotropic conical shell, the dynamic
governing equations of motion can be derived based on Eqs. (4.24)–(4.26). We will now
consider the free vibration of a rotating thin truncated circular isotropic conical shell
under an initial uniform pressure load. The term *uniform* implies a uniform load
distribution, such as an internal or external hydrostatic (or gas) pressure. When the initial
shell is stationary, and if the shell is cylindrical, such an initial uniform pressure load
yields a uniform static stress field in the shell, namely, the stress field does not varying

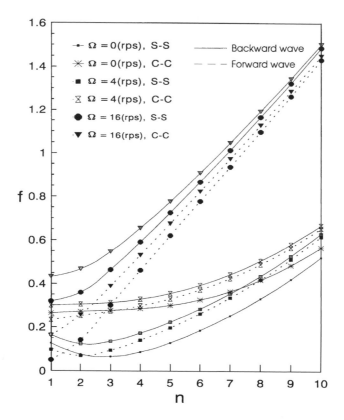

Figure 4.7 Variation of the frequency parameter f with the
circumferential wave number n for cone angle $\alpha = 5°$
$(m = 1, \mu = 0.3, h/a = 0.02, L/a = 20)$.

with the spatial coordinates. However, for a conical shell, the corresponding stress field is a function of spatial coordinates.

In the free vibration of rotating conical shells, the stresses in the shell consist of the initial stress arising from the initial pressure load and additional vibrational stresses such as those due to rotation. Usually, the bending stress in the initial loading state is neglected. The displacements due to the membrane stress are also frequently neglected. These assumptions will be used in the present analysis and hence result in the uncoupling of the initial and vibrational stresses. It is also assumed that the initial stress field is in equilibrium. Therefore, the state of static equilibrium may be taken as a reference state for the deformations due to vibration.

We now consider a rotating conical shell subjected to an initial uniform pressure load that is directionally constant. In other words, the pressure direction does not change as

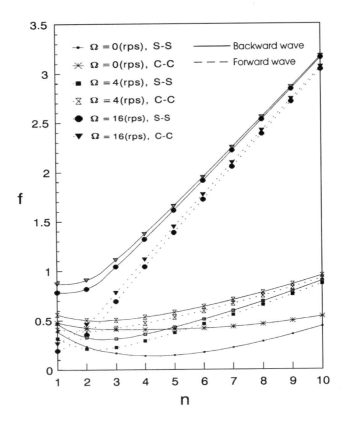

Figure 4.8 Variation of the frequency parameter f with the circumferential wave number n for cone angle $\alpha = 15°$ ($m = 1$, $\mu = 0.3$, $h/a = 0.02$, $L/a = 20$).

the shell deforms during vibration, but remains in its initial direction. Based on the above assumptions and the governing equations (4.24)–(4.26), and using the derivation process set out by Leissa [1993], the governing equations of free vibration for a rotating thin truncated circular isotropic conical shell under an initial uniform pressure load can be derived as

$$\frac{\partial N_x}{\partial x} + \frac{1}{r(x)}\frac{\partial N_{x\theta}}{\partial \theta} + \frac{N_\theta^0}{r^2(x)}\left(\frac{\partial^2 u}{\partial \theta^2} - r(x)\cos\alpha\frac{\partial w}{\partial x}\right)$$

$$+ \frac{\sin\alpha}{r(x)}(N_x - N_\theta) + 2\rho h\Omega \sin\alpha\frac{\partial v}{\partial t} - \rho h\frac{\partial^2 u}{\partial t^2} = 0 \tag{4.27}$$

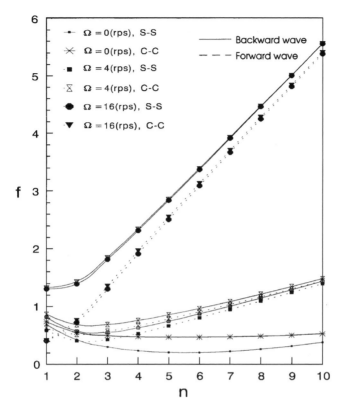

Figure 4.9 Variation of the frequency parameter f with
the circumferential wave number n for cone angle $\alpha = 30°$
$(m = 1,\ \mu = 0.3,\ h/a = 0.02,\ L/a = 20)$.

$$\frac{\partial N_{x\theta}}{\partial x} + \frac{1}{r(x)}\frac{\partial N_\theta}{\partial \theta} + \frac{\cos\alpha}{r(x)}\frac{\partial M_{x\theta}}{\partial x} + \frac{\cos\alpha}{r^2(x)}\frac{\partial M_\theta}{\partial \theta}$$

$$+ \frac{N_\theta^0}{r^2(x)}\left(r(x)\frac{\partial^2 u}{\partial x\,\partial\theta} + \sin\alpha\frac{\partial u}{\partial\theta} + r(x)\sin\alpha\frac{\partial v}{\partial x} - v\sin^2\alpha\right)$$

$$+ 2\frac{\sin\alpha}{r(x)}N_{x\theta} - 2\rho h\Omega\left(\sin\alpha\frac{\partial u}{\partial t} + \cos\alpha\frac{\partial w}{\partial t}\right) - \rho h\frac{\partial^2 v}{\partial t^2} = 0 \qquad (4.28)$$

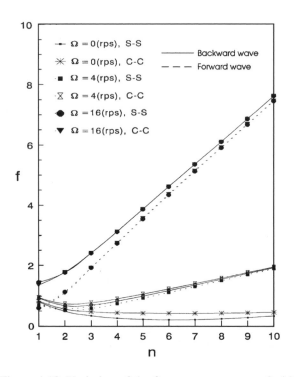

Figure 4.10 Variation of the frequency parameter f with
the circumferential wave number n for cone angle $\alpha = 45°$
($m = 1$, $\mu = 0.3$, $h/a = 0.02$, $L/a = 20$).

$$
\frac{\partial^2 M_x}{\partial x^2} + \frac{2}{r(x)} \frac{\partial^2 M_{x\theta}}{\partial \theta \, \partial x} + \frac{1}{r^2(x)} \frac{\partial^2 M_\theta}{\partial \theta^2} + \frac{2 \sin \alpha}{r(x)} \frac{\partial M_x}{\partial x} - \frac{\sin \alpha}{r(x)} \frac{\partial M_\theta}{\partial x}
$$

$$
+ \frac{N_\theta^0}{r^2(x)} \left(\frac{\partial^2 w}{\partial \theta^2} - r(x) \cos \alpha \frac{\partial u}{\partial x} \right) + \frac{N_\theta^0}{r^2(x)} (w \cos^2 \alpha + u \sin \alpha \cos \alpha)
$$

$$
+ \frac{1}{(1 - \mu^2)} \left[N_x^i \frac{\partial^2 w}{\partial x^2} + N_\theta^i \left(\frac{1}{r^2(x)} \frac{\partial^2 w}{\partial \theta^2} + \frac{\sin \alpha}{r(x)} \frac{\partial w}{\partial x} \right) \right]
$$

$$
- \frac{\cos \alpha}{r(x)} N_\theta + 2\rho h \Omega \cos \alpha \frac{\partial v}{\partial t} - \rho h \frac{\partial^2 w}{\partial t^2} = 0 \tag{4.29}
$$

where N_x^i and N_θ^i are the initial forces in the shell due to the initial uniform pressure load.

It is noted that the above governing equations of motion include the influence of the initial stress field due to the initial uniform pressure load, the effects of initial hoop tension, as well as the centrifugal and Coriolis accelerations arising from the shell rotation.

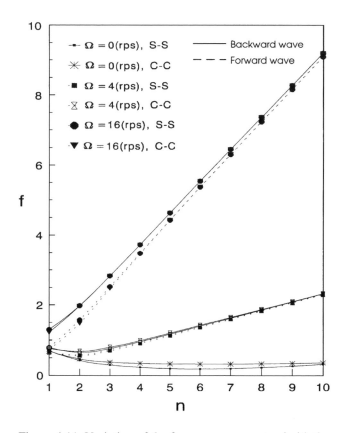

Figure 4.11 Variation of the frequency parameter f with the
circumferential wave number n for cone angle $\alpha = 60°$
$(m = 1, \mu = 0.3, h/a = 0.02, L/a = 20)$.

Table 4.13

Material property of the sandwich-type conical shell.

Layer	E (N/m^2)	μ	ρ (kg/m^3)	Thickness[a]
Inner	4.8265×10^9	0.3	1314	h_{Inner}
Middle	2.0685×10^{11}	0.3	8053	h_{Middle}
Outer	4.8265×10^9	0.3	1314	h_{Outer}

[a] Total shell thickness $h = h_{Inner} + h_{Middle} + h_{Outer}$ and $h_{Inner} = h_{Outer} = h_{Middle}/3$

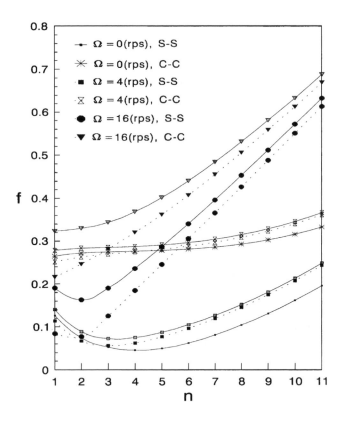

Figure 4.12 Variation of the frequency parameter f with the circumferential wave number n for $\alpha = 5°$ and different rotating velocities Ω ($m = 1$, $h/a = 0.01$, $L/a = 20$).

For a truncated circular conical shell subjected to a uniform pressure load p_0, the following initial static force field in the shell as given by Weingarten [1966] and will be used in the present analysis

$$N_x^i = \frac{p_0}{4} \tan \alpha (a + b) \left(1 + \frac{x \sin \alpha}{a} \right)$$

(4.30)

$$N_\theta^i = \frac{p_0}{2} \tan \alpha (a + b) \left(1 + \frac{x \sin \alpha}{a} \right)$$

(4.31)

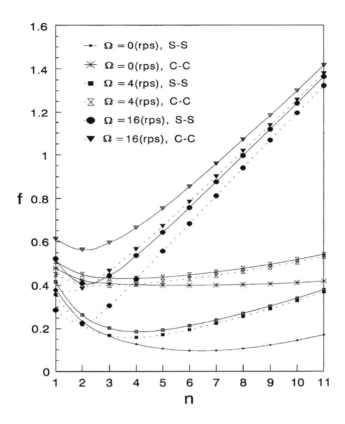

Figure 4.13 Variation of the frequency parameter f with the circumferential wave number n for $\alpha = 15°$ and different rotating velocities Ω ($m = 1$, $h/a = 0.01$, $L/a = 20$).

4.3   Numerical implementation.

a)   Assumed-mode method and generalized differential quadrature.

In this section, an assumed-mode method based on the GDQ for numerical analysis of the free vibration of rotating conical shells, is presented. Before using the assumed-mode method, we need obtain the governing equations in the form where the only variables are the displacements. Specifically, by substituting the geometric deformation relationships (2.15)–(2.17) into the constitutive relationship (2.34), followed by substituting the resulting equations into the governing equations (4.24)–(4.26) defined by the forces, moments and displacements, a set of three-dimensional partial differential governing equations with variable coefficients in terms of the displacements u, v, and w is

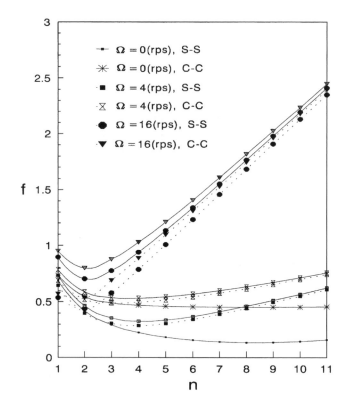

Figure 4.14 Variation of the frequency parameter f with the circumferential wave number n for $\alpha = 30°$ and different rotating velocities Ω ($m = 1$, $h/a = 0.01$, $L/a = 20$).

derived as

$$L_{11}u + L_{12}v + L_{13}w = 0,$$

$$L_{21}u + L_{22}v + L_{23}w = 0, \tag{4.32}$$

$$L_{31}u + L_{32}v + L_{33}w = 0$$

or in vector form as

$$\mathbf{LU} = \mathbf{0} \tag{4.33}$$

where $\mathbf{U}^{\mathrm{T}} = \{u(x, \theta, t), v(x, \theta, t), w(x, \theta, t)\}$ is a displacement field vector of the rotating conical shell. $\mathbf{L} = [L_{ij}]$ ($i, j = 1, 2, 3$) is a 3×3 differential operator matrix of $\mathbf{U}$ and the differential operators L_{ij} ($i, j = 1, 2, 3$) are detailed in Appendix A of this chapter.

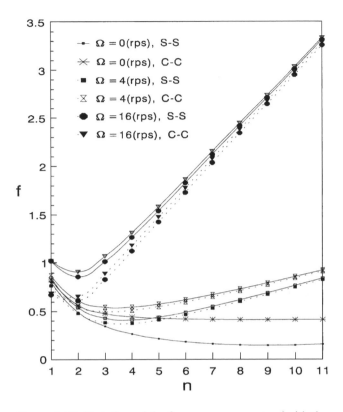

Figure 4.15 Variation of the frequency parameter f with the circumferential wave number n for $\alpha = 45°$ and different rotating velocities Ω ($m = 1$, $h/a = 0.01$, $L/a = 20$).

For the present analysis, the displacement field is assumed to be of the form of a product of unknown continuous smooth functions in the meridional direction and trigonometric functions along the circumferential direction, i.e.,

$$\mathbf{U} = \left\{ \begin{array}{c} u(x, \theta, t) \\ v(x, \theta, t) \\ w(x, \theta, t) \end{array} \right\} = \left\{ \begin{array}{c} U(x)\cos(n\theta + \omega t) \\ V(x)\sin(n\theta + \omega t) \\ W(x)\cos(n\theta + \omega t) \end{array} \right\} \qquad (4.34)$$

where ω (rad/s) is the natural frequency of the rotating conical shell and n is an integer representing the circumferential wave number. The set of partial differential governing equations (4.32) or (4.33) with variable coefficients in the temporal–spatial domain can be transformed mathematically, in a straightforward manner, into a set of ordinary

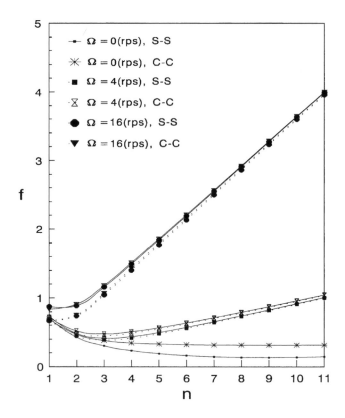

Figure 4.16 Variation of the frequency parameter f with the circumferential wave number n for $\alpha = 60°$ and different rotating velocities Ω ($m = 1$, $h/a = 0.01$, $L/a = 20$).

differential equations with variable coefficients in only the spatial domain of the meridional x-direction

$$L_{11}^* U(x) + L_{12}^* V(x) + L_{13}^* W(x) = 0,$$
$$L_{21}^* U(x) + L_{22}^* V(x) + L_{23}^* W(x) = 0, \qquad (4.35)$$
$$L_{31}^* U(x) + L_{32}^* V(x) + L_{33}^* W(x) = 0$$

or in vector form

$$\mathbf{L}^* \mathbf{U}^* = \mathbf{0} \qquad (4.36)$$

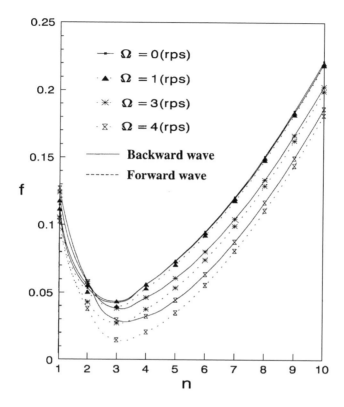

Figure 4.17 Variation of frequency parameter f with circumferential wave number n for a rotating composite conical shell with the Ss–Cl boundary condition ($\alpha = 5°$, $h/a = 0.025$, $L/a = 15$, Rem $= 3$).

where $\mathbf{U}^{*T} = \{U(x), V(x), W(x)\}$ is an unknown modal spatial function vector in the meridional x-direction of the rotating conical shell. $\mathbf{L}^* = [L_{ij}^*]$ $(i, j = 1, 2, 3)$ is a 3×3 differential operator matrix of $\mathbf{U}^*$ and the differential operators L_{ij}^* $(i, j = 1, 2, 3)$ are detailed in Appendix B of this chapter.

In the literature, many numerical approaches have been developed for the solution of differential equations. In general, traditional methods such as the FEM and the finite difference method (FDM) require a large number of grid points for attaining very accurate results. However, for most practical engineering problems, the numerical solution of partial differential equations is required at only a few specified points in a domain. To obtain results at or around a point of interest with acceptable accuracy, these traditional techniques may very often require the use of large numbers of grid

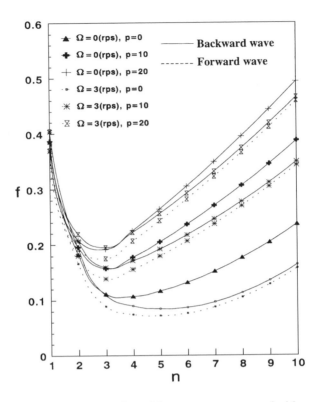

Figure 4.18 Variation of frequency parameter f with circumferential wave number n for a rotating conical shell with the Ss–Cl boundary condition ($\alpha = 15°$, $h/a = 0.025$, $L/a = 15$).

points. The computing resource requirements are thus often unnecessarily large in such cases.

In seeking a more efficient numerical method using fewer grid points to obtain results with acceptable level of accuracy, a global approximate numerical technique, termed the GDQ method [Shu, 1991; Shu & Richard, 1992; Shu *et al.*, 1994; Du *et al.*, 1994, 1995], is developed. It is based on the simple mathematical concept that any sufficiently smooth function in a domain can be expressed approximately as a $(N - 1)$th order polynomial in the overall domain (here N is the number of total discrete mesh points). In other words, at a discrete mesh point in a domain, the derivative of a sufficiently smooth function with respect to a coordinate direction can be approximated by taking a weighted linear sum of the functional values at all the discrete mesh points in that coordinate direction. Taking a one-dimensional function $f(x, t)$ as an example, based

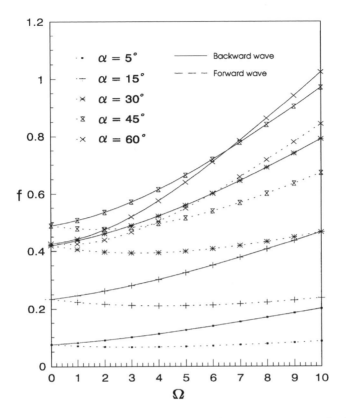

Figure 4.19 Relationship between the frequency parameter f
and rotating speed Ω (rps or Hz) at various cone angles α
($m = 1$, $n = 2$, $\mu = 0.3$, $h/a = 0.01$, $L/a = 15$).

on the above mathematical description of the GDQ method, the mth order derivative of $f(x,t)$ with respect to the coordinate x-direction at the ith discrete mesh point $x = x_i$ should satisfy the following linear constrained relation

$$\frac{\partial^m f(x,t)}{\partial x^m}\bigg|_{x=x_i} = \sum_{j=1}^{N} C_{ij}^m f(x_j,t), \qquad i = 1,2,...,N \tag{4.37}$$

where N is the number of total discrete grid points in the x-direction used in the approximation process, and C_{ij}^m is the weighting coefficient associated with the mth order derivative. In order to construct the weighting coefficient C_{ij}^m, the present GDQ method

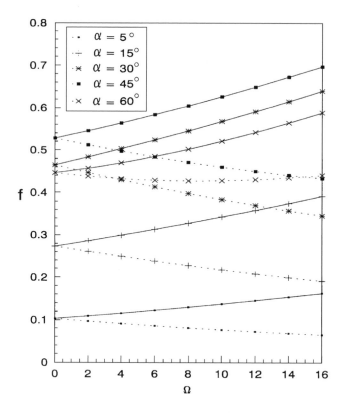

Figure 4.20 Relationship between the frequency parameter f
and rotating speed Ω (rps or Hz) at various cone angles α for
an orthotropic conical shell ($m = 1$, $n = 1$, $h/a = 0.02$,
$L/a = 15$, Rem = 3).

takes the following Lagrange interpolated polynomial as a trial function of Eq. (4.37)

$$g_k^*(x) = \frac{M(x)}{(x - x_k)M^{(1)}(x_k)}, \qquad k = 1, 2, ..., N \tag{4.38}$$

where N is the number of total discrete mesh points, x_k ($k = 1, 2, ..., N$) is the coordinate of the discrete grid points which can be chosen arbitrarily; $M^{(1)}(x)$ is the first derivative of $M(x)$ and may be defined as follows:

$$M(x) = \prod_{j=1}^{N} (x - x_j), \qquad M^{(1)}(x_k) = \prod_{j=1(j \neq k)}^{N} (x_k - x_j) \tag{4.39}$$

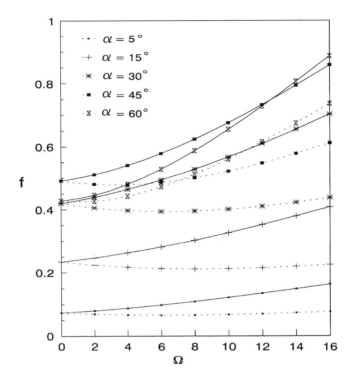

Figure 4.21 Relationship between the frequency parameter f
and rotating speed Ω (rps or Hz) at various cone angles α for
a sandwich-type conical shell ($m = 1$, $n = 2$, $h/a = 0.008$,
$L/a = 20$).

Following derivation, the weighting coefficients for the first derivative are
obtained as

$$C_{ij}^1 = \frac{M^{(1)}(x_i)}{(x_i - x_j)M^{(1)}(x_j)}, \qquad \text{for } i \neq j \text{ and } i,j = 1,2,...,N \tag{4.40}$$

and

$$C_{ii}^1 = -\sum_{j=1(j\neq i)}^{N} C_{ij}^1, \qquad \text{for } i = 1,2,...,N \tag{4.41}$$

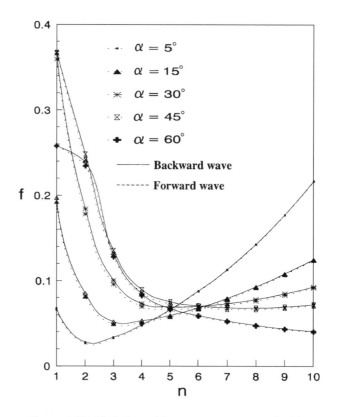

Figure 4.22 Variation of frequency parameter f with
circumferential wave number n for a rotating composite
conical shell with Ss–Sl boundary condition
($h/a = 0.03$, $L/a = 20$, $\Omega = 0.3$ rps, Rem = 3).

For the second and higher order derivatives, the weighting coefficients are obtained
by the following simple recurrence relationship

$$C_{ij}^m = m\left(C_{ij}^1 C_{ii}^{m-1} - \frac{C_{ij}^{m-1}}{x_i - x_j} \right),$$
(4.42)

for $i \neq j$ and $i, j = 1, 2, ..., N$; $m = 2, 3, ..., N - 1$

$$C_{ii}^m = -\sum_{j=1(j \neq i)}^{N} C_{ij}^m \qquad \text{for } i = 1, 2, ..., N$$
(4.43)

It can be observed in the derivation procedure of the GDQ method that, as an
improved differential quadrature (DQ) method, the GDQ method has two advantages
over the DQ method [Bellman & Casti, 1971; Bellman *et al.*, 1972; Jan *et al.*, 1989;

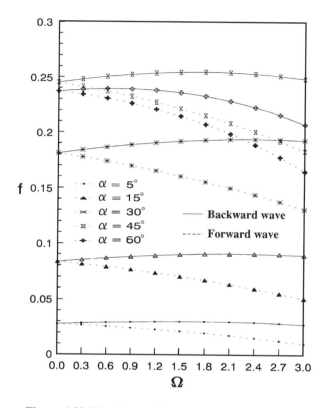

Figure 4.23 Variation of frequency parameter f with
rotating speed Ω for a rotating composite conical shell
with the Ss–Sl boundary condition ($h/a = 0.03$,
$L/a = 20$, $n = 2$, Rem = 3).

Quan & Chang, 1989; Lam, 1993; Wang & Bert, 1993; Kim & Bert, 1993; Bert *et al.*, 1993, 1994; Gutierrez *et al.*, 1994; Gutierrez & Laura, 1995; Striz *et al.*, 1995; Malik & Bert, 1995, 1996; Bert & Malik, 1996a-f; Kukreti *et al.*, 1996; Liew *et al.*, 1996]. The first is that the weighting coefficients C_{ij}^m can be determined by a simple recurrence relationship instead of having to solve a set of linear algebraic equations. The second is that there is no restriction on the number and coordinate distribution of discrete mesh points. Therefore, for the rotating conical shells, the discrete mesh points in the meridional x-direction can be chosen arbitrarily. In the present application of the GDQ method, however, only two distributions of the discrete mesh points in the meridional x-direction are considered:

1)   a uniform equidistant distribution of discrete mesh points

$$x_i = \frac{i-1}{N-1}L, \qquad i = 1, 2, ..., N \tag{4.44}$$

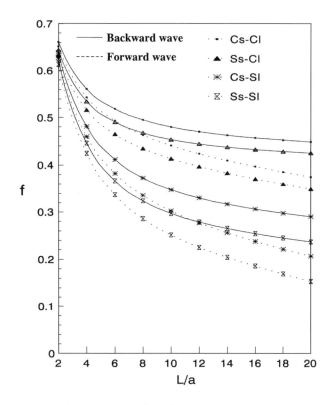

Figure 4.24 Variation of the frequency parameter f with
the length ratio L/a ($\alpha = 30°$, $h/a = 0.005$, $n = 2$,
Cone III, $\Omega = 4$ rps).

2)   a cosine distribution of discrete mesh points, in which a denser distribution of the
discrete mesh points is given near the edges

$$x_i = \frac{1}{2}\left(1 - \cos\left(\frac{i-1}{N-1}\pi\right)\right)L \qquad i = 1, 2, ..., N \tag{4.45}$$

It is noted that both the above distributions include the discrete mesh points at both
edges of the shell, i.e., at $x_1 = 0$ and $x_N = L$.

In the present application of the GDQ method, four boundary conditions for rotating
conical shell are considered, namely,

Table 4.14

Layered configuration and material property of the three sandwich-type conical shells.

Cone	Layer	Thickness[a]	E (N/m^2)	μ	ρ (kg/m^3)
I	Inner	$h/5$	4.8265×10^9	0.3	1314
	Middle	$3h/5$	2.0685×10^{11}	0.3	8053
	Outer	$h/5$	4.8265×10^9	0.3	1314
II	Inner	$3h/7$	4.8265×10^9	0.3	1314
	Middle	$h/7$	2.0685×10^{11}	0.3	8053
	Outer	$3h/7$	4.8265×10^9	0.3	1314
III	Inner	$h/3$	4.8265×10^9	0.3	1314
	Middle	$h/3$	2.0685×10^{11}	0.3	8053
	Outer	$h/3$	4.8265×10^9	0.3	1314

[a] h is the total thickness of the sandwich-type conical shell

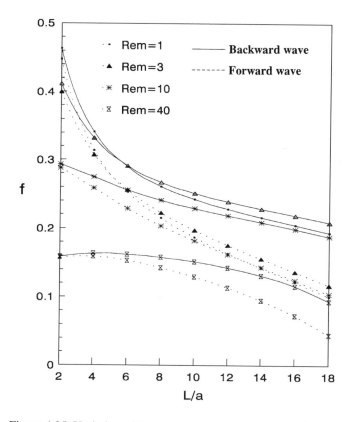

Figure 4.25 Variation of frequency parameter f with length ratio
L/a for a rotating composite conical shell with the Ss–Sl boundary
condition ($\alpha = 30°$, $h/a = 0.005$, $\Omega = 5$ rps, $n = 2$).

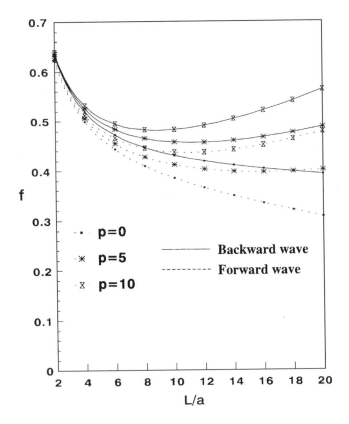

Figure 4.26 Variation of frequency parameter f with length ratio L/a for a rotating conical shell with the Cs–Cl boundary condition ($\alpha = 30°$, $h/a = 0.01$, $\Omega = 5$ rps, $n = 2$).

1)   clamped at small edge-clamped at large edge (Cs–Cl or C–C),

$$u = 0, \qquad v = 0, \qquad w = 0, \qquad \frac{\partial w}{\partial x} = 0 \qquad \text{at } x = 0 \text{ and } L \qquad (4.46)$$

2)   simply supported at small edge-clamped at large edge (Ss–Cl)

$$v = 0, \qquad w = 0, \qquad N_x = 0, \qquad M_x = 0 \qquad \text{at } x = 0,$$

$$u = 0, \qquad v = 0, \qquad w = 0, \qquad \frac{\partial w}{\partial x} = 0, \qquad \text{at } x = L \qquad (4.47)$$

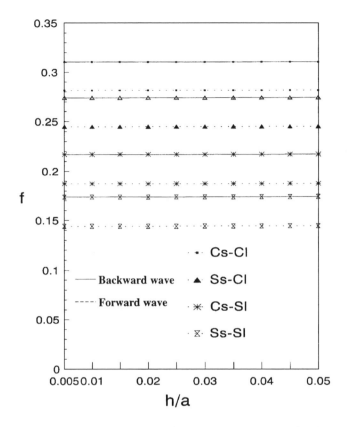

Figure 4.27 Variation of the frequency parameter f with the
thickness ratio h/a ($\alpha = 15°$, $L/a = 8$, $n = 2$, Cone II,
$\Omega = 4$ rps).

3) clamped at small edge-simply supported at large edge (Cs–Sl)

$$u = 0, \quad v = 0, \quad w = 0, \quad \frac{\partial w}{\partial x} = 0, \quad \text{at } x = 0,$$
$$v = 0, \quad w = 0, \quad N_x = 0, \quad M_x = 0 \quad \text{at } x = L$$

$$(4.48)$$

4) simply supported at small edge-simply supported at large edge (Ss–Sl, or S–S)

$$v = 0, \quad w = 0, \quad N_x = 0, \quad M_x = 0 \quad \text{at } x = 0 \text{ and } L$$

$$(4.49)$$

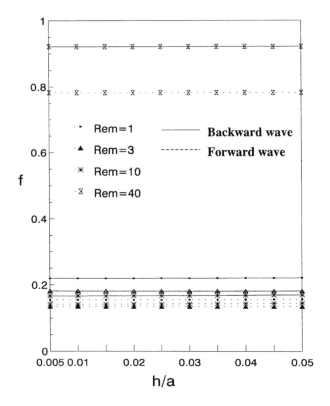

Figure 4.28 Variation of frequency parameter f with thickness ratio h/a for a rotating composite conical shell with the Ss–Sl boundary condition ($\alpha = 60°$, $L/a = 10$, $\Omega = 8$ rps, $n = 2$).

To apply the GDQ method for the analysis of the frequency characteristics of rotating conical shells, one imposes the mathematical expression (4.37), which is the basic concept of the GDQ method, on the set of ordinary differential governing equations (4.35) for rotating conical shells, in terms of unknown modal spatial functions. A set of numerical discrete approximate governing equations is derived in the form of linear algebraic simultaneous equations with reference to the unknown function vector $\mathbf{U}^*$. It can be written in the following matrix form:

$$\mathbf{L}^*\mathbf{U}^*|_{x=x_i} = \mathbf{R}_{3\times 15}\mathbf{U}^{**}_{15\times 1}|_{x=x_i} = \mathbf{0}, \qquad i = 1, 2, 3, ..., N \tag{4.50}$$

where $\mathbf{U}^{*T} = \{U(x), V(x), W(x)\}$ is an unknown modal spatial function vector, and it describes the distribution of the modal amplitudes in the meridional x-direction.

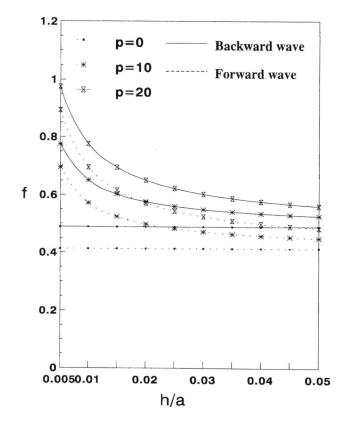

Figure 4.29 Variation of frequency parameter f with thickness ratio h/a for a rotating conical shell with the Cs–Cl boundary condition ($\alpha = 45°, L/a = 10, \Omega = 10$ rps, $n = 2$).

N is the number of total discrete mesh points including the points at both edges. $\mathbf{L}^* = [L_{ij}^*]$ ($i, j = 1, 2, 3$) is a 3×3 differential operator matrix of $\mathbf{U}^*$, as shown in Eqs. (B.1)–(B.9). $\mathbf{R}$ is a 3×15 variable coefficient matrix related to discrete mesh point $x = x_i$. $\mathbf{U}^{**}$ is a 15th order column vector and can be expressed as follows:

$$
\begin{aligned}
\mathbf{U}^{**T}\big|_{x=x_i} = \{ &U(x_i), U^{(1)}(x_i), U^{(2)}(x_i), U^{(3)}(x_i), U^{(4)}(x_i), \\
&V(x_i), V^{(1)}(x_i), V^{(2)}(x_i), V^{(3)}(x_i), V^{(4)}(x_i), \\
&W(x_i), W^{(1)}(x_i), W^{(2)}(x_i), W^{(3)}(x_i), W^{(4)}(x_i) \}
\end{aligned}
\tag{4.51}
$$

where

$$
U^{(m)}(x_i) = \sum_{j=1}^{N} C_{ij}^m U(x_j), \quad V^{(m)}(x_i) = \sum_{j=1}^{N} C_{ij}^m V(x_j), \quad W^{(m)}(x_i) = \sum_{j=1}^{N} C_{ij}^m W(x_j), \quad m = 1, 2, 3, 4.
$$

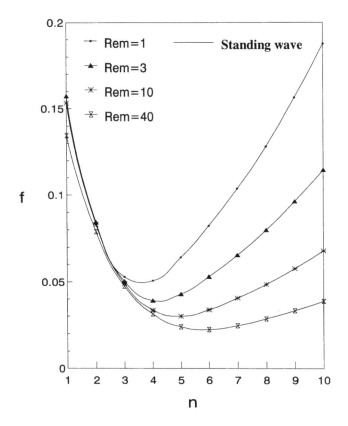

Figure 4.30 Variation of frequency parameter f with circumferential wave number n for a stationary composite shell with Cs–Cl boundary condition ($\alpha = 5°$, $h/a = 0.01$, $L/a = 10$, $\Omega = 0$).

Usually, there are a total of eight boundary conditions at both edges of the rotating shell. For a set of given boundary conditions, by imposing the numerical discrete approximate governing equations (4.50) on every discrete mesh point $x=x_i$ ($i=1$, 2,...,N) in the meridional x-direction, followed by rearranging the resulting expression according to the natural frequency ω, a set of numerical discrete eigenvalue equations with respect to ω is derived and can be written in the following matrix form:

$$[\omega^2 \mathbf{H}_1 + \omega \mathbf{H}_2 + \mathbf{H}_3]\mathbf{d} = \mathbf{0} \qquad (4.52)$$

where $\mathbf{H}_1$, $\mathbf{H}_2$ and $\mathbf{H}_3$ are the $N^* \times N^*$ numerical coefficient matrices ($N^* = 3N - 8$).

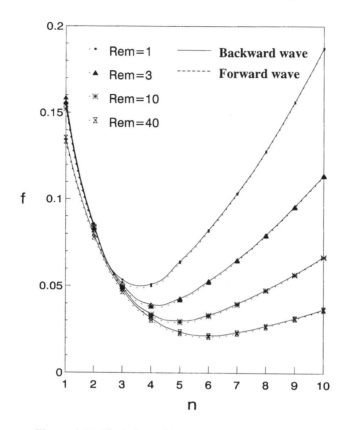

Figure 4.31 Variation of frequency parameter f with
circumferential wave number n for a rotating composite
shell with Cs–Cl boundary condition ($\alpha = 5°$, $h/a = 0.01$,
$L/a = 10$, $\Omega = 0.5$ rps).

$\mathbf{d}$ is a N^*-order modal column vector at the discrete mesh points and it can be written as

$$\mathbf{d}^T = \{U(x_2), U(x_3),..., U(x_{N-2}), U(x_{N-1}),$$
$$V(x_2), V(x_3),..., V(x_{N-2}), V(x_{N-1}),$$
$$W(x_3), W(x_4),..., W(x_{N-3}), W(x_{N-2})\} \tag{4.53}$$

It is evident that the set of numerical discrete eigenvalue equations (4.52) is a non-standard eigenvalue equation. However, it can be transformed into a standard form of eigenvalue equation [Dong, 1977] as

$$\left(\begin{bmatrix} \mathbf{0} & \mathbf{I} \\ -\mathbf{H}_3 & -\mathbf{H}_2 \end{bmatrix} - \omega \begin{bmatrix} \mathbf{I} & \mathbf{0} \\ \mathbf{0} & \mathbf{H}_1 \end{bmatrix} \right) \left\{ \begin{matrix} \mathbf{d} \\ \omega\mathbf{d} \end{matrix} \right\} = \mathbf{0} \tag{4.54}$$

where $\mathbf{I}$ is a $N^* \times N^*$ identity matrix.

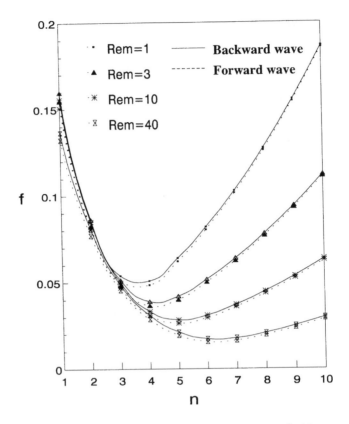

Figure 4.32 Variation of frequency parameter f with
circumferential wave number n for a rotating composite
shell with Cs–Cl boundary condition ($\alpha = 5°$, $h/a = 0.01$,
$L/a = 10$, $\Omega = 1$ rps).

Using a conventional eigenvalue approach, the standard eigenvalue equation (4.54) can be solved and $2N^*$ eigenvalues ω_i ($i = 1, 2, ..., 2N^*$) are obtained. From these $2N^*$ eigenvalues, the two real eigenvalues, one positive and the other negative, whose absolute values are the smallest, are chosen. These two eigenvalues are the eigensolutions, and they correspond, respectively, to the backward and forward travelling waves.

The frequency parameter of the free vibration solution of a rotating conical shell is a function of the rotating speed. At a given rotating speed, the eigensolution for each mode of the vibration, namely, for each pair of the wave numbers (m, n), where m is the meridional wave number and n is the circumferential wave number, consists of a positive and a negative eigenvalue. These two eigenvalues correspond to the backward and forward travelling waves, respectively, or to the positive and negative rotating speeds, respectively. The positive eigenvalue corresponds to the backward wave due to a rotation

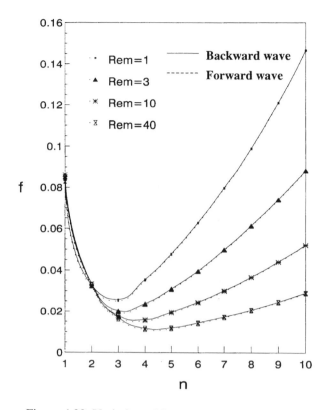

Figure 4.33 Variation of frequency parameter f with
circumferential wave number n for a rotating conical
shell with Ss–Sl boundary condition ($\alpha = 5°$,
$h/a = 0.01$, $L/a = 15$, $\Omega = 0.2$ rps).

$\Omega > 0$; and the negative eigenvalue corresponds to the forward wave due to a rotation $\Omega < 0$. In the case of a stationary conical shell, these two eigenvalues are identical and the vibration of the conical shell is a standing wave motion. However, when the conical shell starts to rotate, the standing wave motion will be transformed, and depending on the rotating direction, backward or forward waves will emerge. Through the generated numerical results, it is observed that the absolute values of the backward waves are generally higher than those of the forward waves.

For ease of the present discussions, a non-dimensional frequency parameter f for a rotating laminated composite conical shell is defined as

$$f = \omega b \sqrt{\frac{\rho h}{A_{11}}} \tag{4.55}$$

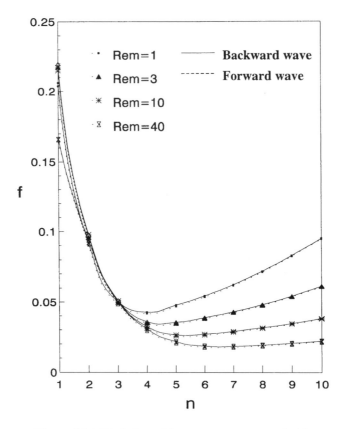

Figure 4.34 Variation of frequency parameter f with
circumferential wave number n for a rotating conical shell
with Ss–Sl boundary condition ($\alpha = 15°$, $h/a = 0.01$,
$L/a = 15$, $\Omega = 0.2$ rps).

If the laminated composite conical shell is reduced to an orthotropic conical shell, the non-dimensional frequency parameter f can be rewritten as follows:

$$f = \omega b \sqrt{\frac{(1 - \mu_{x\theta}\mu_{\theta x})\rho}{E_x}} \tag{4.56}$$

When the laminated composite conical shell is further reduced to an isotropic conical shell, the non-dimensional frequency parameter f becomes

$$f = \omega b \sqrt{\frac{(1 - \mu^2)\rho}{E}} \tag{4.57}$$

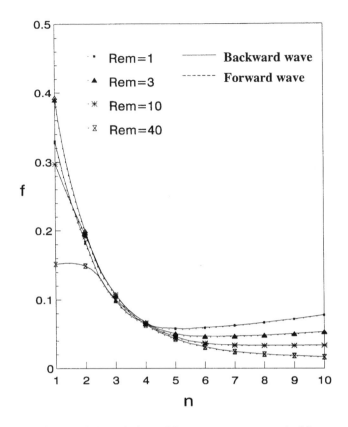

Figure 4.35 Variation of frequency parameter f with
circumferential wave number n for a rotating conical shell
with Ss–Sl boundary condition ($\alpha = 30°$, $h/a = 0.01$,
$L/a = 15$, $\Omega = 0.2$ rps).

It should be noted that the frequency parameter f of the vibration solution of a rotating shell is a function of the rotating speed Ω.

Moreover, if E_x and E_θ denote the elastic moduli, respectively, in the meridional x and circumferential θ-directions for orthotropic single-layer or regular symmetrical cross-ply laminated conical shells, a non-dimensional orthotropic parameter E^* (or "Rem") is defined in the following form:

$$E^* = \text{Rem} = \frac{E_x}{E_\theta} \tag{4.58}$$

Additionally, for ease of presentation on the influence of initial uniform pressure load on frequency characteristics of rotating conical shell, a non-dimensional initial

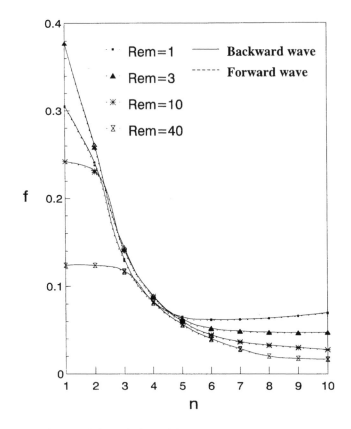

Figure 4.36 Variation of frequency parameter f with circumferential wave number n for a rotating conical shell with Ss–Sl boundary condition ($\alpha = 45°$, $h/a = 0.01$, $L/a = 15$, $\Omega = 0.2$ rps).

pressure parameter p is introduced

$$p = \frac{\text{initial uniform pressure load } (p_0)}{\text{a standard atmosphere}} \qquad (4.59)$$

In the presentation of computed results shown in the subsequent figures, the backward wave is represented by a solid line and the forward wave by a dashed line; the unit of rotating speed Ω is rps (rps, revolutions per second or Hz). In addition, four boundary conditions are considered here for the rotating conical shell. These are the fully clamped (Cs–Cl), simply supported at small edge-clamped at large edge (Ss–Cl), clamped at small edge-simply supported at large edge (Cs–Sl) and fully simply supported (Ss–Sl) boundary conditions.

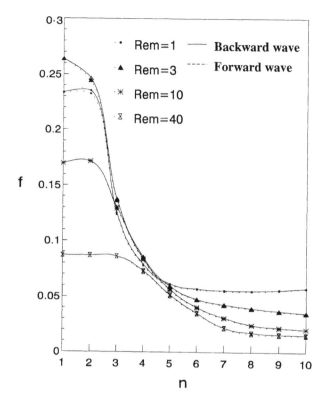

Figure 4.37 Variation of frequency parameter f with circumferential wave number n for a rotating conical shell with Ss–Sl boundary condition ($\alpha = 60°$, $h/a = 0.01$, $L/a = 15$, $\Omega = 0.2$ rps).

In the following section, numerical comparisons are made to validate the GDQ method as a powerful numerical technique, well suited to solve the present rotating conical shell problem. This global approximate method is especially suitable for the solutions of global characteristics such as free vibration or buckling analyses. The numerical accuracy of the GDQ method, with its good weighting characteristics, is highly reliable, and its implementation is simple and efficient. The global functional interpolation prescribes simple explicit formulae for the coefficients making it very convenient for code implementation.

b)   *Convergence characteristics and numerical validation.*

To determine the computational accuracy, the present assumed-mode method based on the GDQ method is validated through the examination of the frequency characteristics

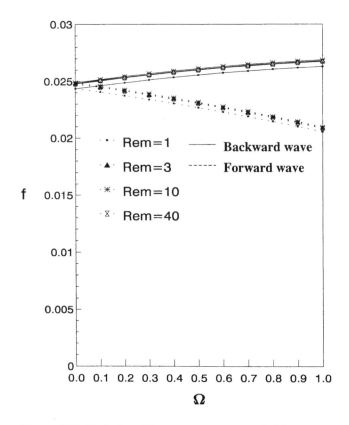

Figure 4.38   Variation of frequency parameter *f* with rotating speed Ω for a rotating composite conical shell with Ss–Sl boundary condition ($\alpha = 5°$, $h/a = 0.01$, $L/a = 20$, $n = 2$).

of various shell configurations. The validation is carried out by comparing the present computed results with those available in the literature. The numerical comparisons of frequency parameters are shown, respectively, for a stationary cylindrical shell, a rotating cylindrical shell, and a stationary conical shell. The comparison for a rotating conical shell is not presented because numerical results for this class of problems are not available in prior literature. This is so because the only prior work on this subject by Sivadas [1995] only reported limited graphical results and did document any numerical data.

For a stationary cylindrical shell, the validation study of the present assumed-mode method based on the GDQ is tabulated in Tables 4.1–4.3. As shown in Table 4.1 for a non-rotating thin isotropic cylindrical shell with simply supported–simply supported boundary conditions, by taking cone angle $\alpha = 0$ and rotating speed $\Omega = 0$, the comparison of frequency parameters is made with Markus [1988], whose results are based

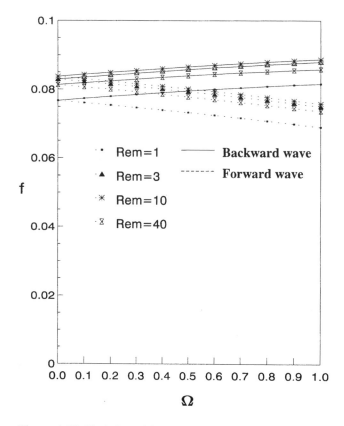

Figure 4.39 Variation of frequency parameter f with rotating
speed Ω for a rotating composite conical shell with Ss–Sl
boundary condition ($\alpha = 15°, h/a = 0.01, L/a = 20, n = 2$).

on three-dimensional elasticity theory. For the convergence study of the GDQ
implementation, this table compares the results of Markus [1988] with that of the
GDQ method using both the cosine and uniform equidistant distributions of the discrete
mesh points in the meridional direction. From the table it is evident that the numerical
GDQ results converge rapidly with increasing number of discrete mesh points. Generally
good agreement with Markus [1988] can be observed. For a non-rotating thin isotropic
clamped–clamped cylindrical shell, the numerical results of the frequency parameters
from the present GDQ methodology are, respectively, compared with the exact solutions
of Dym [1973] and the series solutions of Chung [1981], in Table 4.2. From the table, it
can be seen that the numerical solution of the GDQ method, employing only a few
discrete mesh points, is equivalent to the series solution using a large number of terms.
Again, a very good agreement in the comparisons is observed. For a non-rotating thin

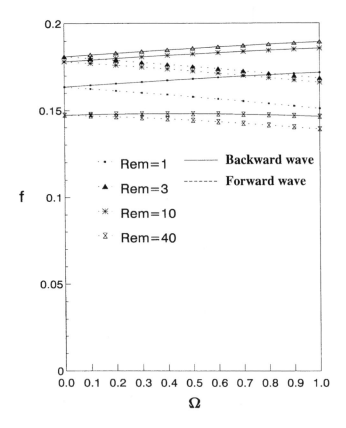

Figure 4.40 Variation of frequency parameter *f* with rotating
speed Ω for a rotating composite conical shell with Ss–Sl
boundary condition ($\alpha = 30°, h/a = 0.01, L/a = 20, n = 2$).

isotropic circular clamped–clamped cylindrical shell subjected to an initial internal uniform pressure load, Table 4.3 presents the numerical comparison between the presently generated GDQ numerical results with the experimentally measured frequencies (cps) of Miserentino & Vosteen [1965], which appeared in a NASA report. The present GDQ results are obtained by taking NGP = 10 (here, NGP refers to the total number of discrete mesh points in the shell meridional direction). The isotropic cylindrical shells used in the experiment are made of various metals including 17–7 PH stainless steel, 2024 aluminum and 301 stainless steel. From the table, it is noted that the maximum relative discrepancy between experimental and theoretical results is less than 10%. Considering the possible limitations in the experimental analysis, the agreement shown in the table is satisfactory.

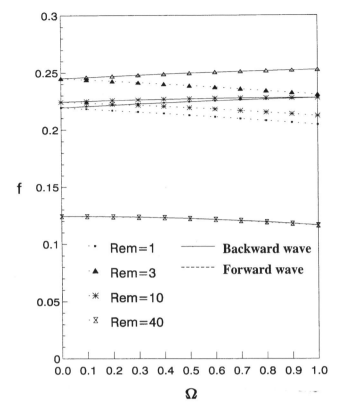

Figure 4.41 Variation of frequency parameter f with rotating
speed Ω for a rotating composite conical shell with Ss–Sl
boundary condition ($\alpha = 45°, h/a = 0.01, L/a = 20, n = 2$).

For a rotating cylindrical shell, results in Tables 4.4–4.6 are generated to validate the present GDQ-based assumed-mode methodology. All comparisons of the frequency parameters for both backward and forward waves are made with solutions generated via a solution equation derived by Chen *et al.* [1993], for an infinitely long rotating cylindrical shell. Table 4.4 shows the comparison of the frequency parameter of the backward waves, while Table 4.5 presents the corresponding comparison for the forward waves. Both tables reveal the good convergence characteristics of the GDQ method with increasing discrete mesh points, as well as the very good agreement with the results of Chen *et al.* [1993]. Table 4.6 shows the comparison of present results with those of Lam & Loy [1995a], and Chen *et al.* [1993]. Lam & Loy [1995a] used trigonometric functions as the displacement field trial functions, and obtained the eigensolutions directly from the governing equations of motion. Once again, very good correspondence is observed in the comparison.

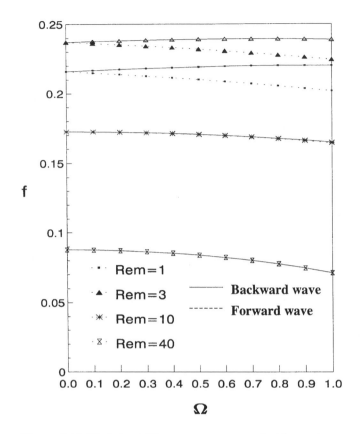

Figure 4.42 Variation of frequency parameter f with rotating speed Ω for a rotating composite conical shell with Ss–Sl boundary condition ($\alpha = 60°, h/a = 0.01, L/a = 20, n = 2$).

To validate the present GDQ-based assumed-mode methodology for the non-rotating isotropic conical shell with different boundary conditions and various geometric properties, numerical comparisons of the frequency parameters are made with the numerical integral results of Irie *et al.* [1984], and these are tabulated in Tables 4.7–4.11. The results compared in Table 4.7 demonstrate the good stable convergence characteristics of the GDQ method. Also, very good agreements are observed for non-rotating conical shells with respective clamped (Cs–Cl) and simply supported (Ss–Sl) boundary conditions. To conduct a thorough numerical comparison for conical shells with different shell parameters, especially for different boundary conditions, Tables 4.8–4.11 are tabulated. Table 4.8 considers conical shells with Cs–Cl boundary conditions, Table 4.9 the corresponding Ss–Sl case, and Tables 4.10 and 4.11 the corresponding Ss–Cl and Cs–Sl cases, respectively. For all the different boundary

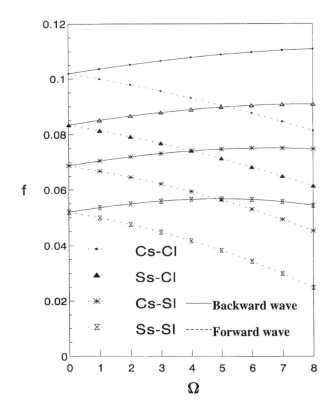

Figure 4.43 Variation of the frequency parameter f with
rotating speed Ω ($\alpha = 5°$, $h/a = 0.015$, $L/a = 10$,
$n = 2$, Cone I).

condition cases, the comparisons in these tables indicate that the present GDQ results are generally in good agreement with those reported by Irie *et al.* [1984].

The final comparison in this section is for a non-rotating truncated circular isotropic aluminum conical shell with the Ss–Sl boundary condition, and subjected to an initial external uniform pressure. Results are tabulated in Table 4.12, and comparisons made with both the experimental and theoretical results reported by Weingarten [1966]. In the theoretical analysis of Weingarten [1966], the meridional and circumferential displacements u and v were assumed to be dependent on the normal displacement w. As a result, the governing equations of motion were first simplified to a set of uncoupled equations in terms of w, and the Galerkin method was then used to extract the frequencies. From the table, it is evident that, when compared with the experimental data, the present theoretical results by the GDQ method are more refined over those presented by Weingarten [1966]. Furthermore, when experimental limitations are considered, the agreement shown in the table may be

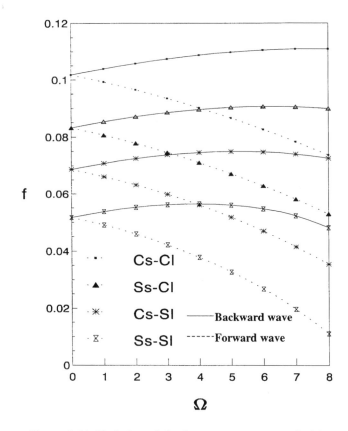

Figure 4.44 Variation of the frequency parameter f with rotating speed Ω ($\alpha = 5°$, $h/a = 0.015$, $L/a = 10$, $n = 2$, Cone II).

regarded as very good, as it can be difficult to accurately impose the desired conditions in an experimental environment. One of the major sources of experimental error results from the imposition of boundary conditions, where they are very often too rigidly imposed, especially simply supported boundary condition. It can be very difficult to allow bending moment and yet restrict the transverse displacement.

4.4   Frequency characteristics.

a)   *Influence of rotating velocity.*

As mentioned earlier, in the present discussion, the rotating angular velocity of the shell of revolution is the differentiating effect between the stationary and rotating shells.

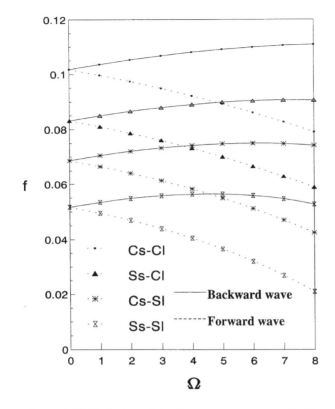

Figure 4.45 Variation of the frequency parameter f with
rotating speed Ω ($\alpha = 5°$, $h/a = 0.015$, $L/a = 10$,
$n = 2$, Cone III).

Arising from this constant rotation there are the Coriolis and centrifugal accelerations as well as hoop tension which significantly affect the dynamic behaviors of the shell. In this section, the influence of rotating velocity on the free vibration of truncated circular conical shells is examined in detail.

Figures 4.2–4.6 present the influence of rotating velocity Ω on the frequency characteristics of conical shells with simply supported boundary conditions at both ends. The relationship between the frequency parameter f and the circumferential wave number n are presented for five different cone angles, $\alpha = 5, 15, 30, 45$ and $60°$ and at five different rotating speeds, $\Omega = 0, 1, 4, 8$ and 16 rps. From the figures, it can be seen that, for the case of a stationary cone, $\Omega = 0$, a standing wave is obtained, and for other values of Ω, backward and forward waves are obtained. From the figures, it can also be observed that the frequency parameter f generally increases with increasing rotating speed Ω for the same circumferential wave number n.

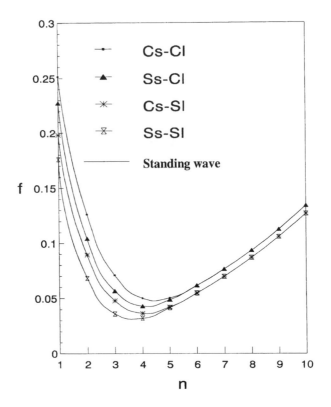

Figure 4.46 Variation of frequency parameter f with circumferential wave number n for a stationary conical shell ($\alpha = 5°$, $h/a = 0.01$, $L/a = 8$, Cone I, $\Omega = 0$).

To examine the coupled influence of the rotating velocity Ω and boundary condition on the frequency characteristics, Figs. 4.7–4.11 are generated. These figures depict the variations of the frequency parameter f with the circumferential wave number n for the rotating conical shells having C–C and S–S boundary conditions, with geometric properties $h/a = 0.02$ and $L/a = 20$. These five figures correspond to the five different cone angles, namely, $\alpha = 5, 15, 30, 45$ and $60°$ and cover three different rotating velocities, $\Omega = 0, 4$ and 16 rps. For a corresponding sandwich-type conical shell with properties tabulated in Table 4.13, Figs. 4.12–4.16 are generated for similar variations of the frequency parameter f.

For the case of the non-rotating conical shell, i.e., $\Omega = 0$, it can be seen from Figs. 4.7–4.11 that there are significant differences between the frequency parameters corresponding to the C–C and S–S boundary conditions. These differences are observed to decrease with increasing cone angles. For the rotating conical shell cases, similar

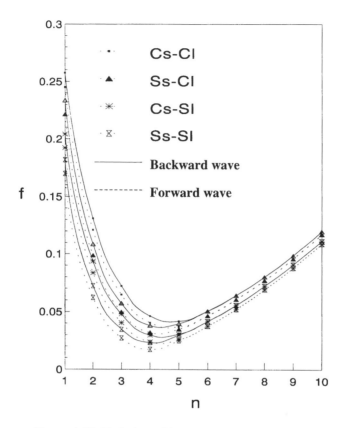

Figure 4.47 Variation of frequency parameter f with circumferential wave number n ($\alpha = 5°$, $h/a = 0.01$, $L/a = 8$, Cone I, $\Omega = 3$ rps).

observations are also noted. The frequency parameter f increases with increase in the rotating velocity Ω for a given circumferential wave number n. It should also be noted that, as the circumferential wave number n increases to high values, the difference in the frequency parameters between the C–C and S–S boundary conditions decreases for a given rotating velocity Ω. This difference also decreases with increase in the rotating velocity Ω for a given circumferential wave number n. However, the frequency curve corresponding to the C–C boundary condition is always larger than that corresponding to the S–S boundary condition. Therefore, we can conclude that boundary conditions do influence the relationship between the frequency parameter f and circumferential wave number n. This influence is more significant for smaller cone angles, lower circumferential wave numbers n, and slower rotating velocities. However, as the circumferential wave number n or the cone angle α increases, such influences tend

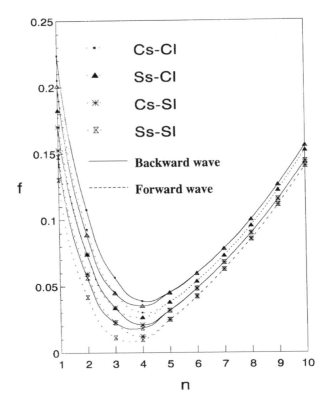

Figure 4.48 Variation of frequency parameter f with circumferential wave number n ($\alpha = 5°$, $h/a = 0.015$, $L/a = 10$, Cone I, $\Omega = 4$ rps).

to become insignificant for a given rotating velocity, and the influence of rotating speed Ω is now much more significant than that of the boundary condition.

For the rotating glass fibre–epoxy laminated composite conical shell with cone angle $\alpha = 5°$, Fig. 4.17 presents the influence of rotating speed Ω on the relationship between frequency parameter f and circumferential wave number n. This rotating composite conical shell is simply supported at the small edge and clamped at the large edge (Ss–Cl). From the figure it can be observed that the influence of rotating speed Ω at large circumferential wave number n is more significant than that at small circumferential wave number n. The difference in the frequency parameters f between backward and forward waves increases with increasing rotating speed Ω.

When initial uniform pressure is applied on a rotating Ss–Cl isotropic conical shell with cone angle $\alpha = 15°$, Fig. 4.18 shows the influence of rotating speed Ω on the variation of frequency parameter f with circumferential wave number n for three different

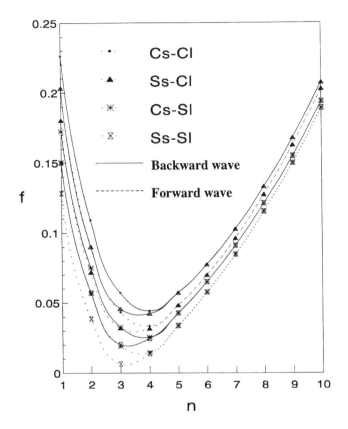

Figure 4.49 Variation of frequency parameter f with circumferential wave number n ($\alpha = 5°$, $h/a = 0.02$, $L/a = 10$, Cone I, $\Omega = 5$ rps).

initial pressure parameters p. From Fig. 4.18, it can be seen that with increasing circumferential wave number n, the frequency parameters f of both the backward and forward waves first decrease and then increase. The gradients of the decreasing frequency parameters f are almost the same for different initial pressure parameters p and rotating speeds Ω. However, the gradients of the increasing frequency parameters f are different depending on the initial pressure parameter p and rotating speed Ω. Meanwhile, it can also be seen that the influence of initial pressure parameter p is more significant than that of rotating speed Ω when circumferential wave number n is large. Further, the influence of the rotating speed Ω at large circumferential wave number n is more significant than that at small circumferential wave number n. Finally, we also note that the difference in frequency parameters f between backward and forward waves increases with increasing rotating speed Ω.

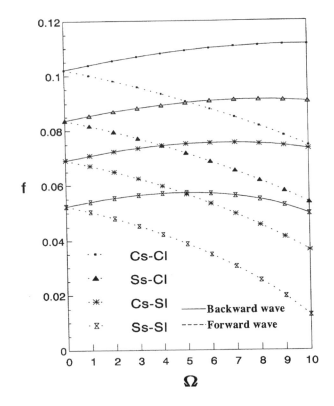

Figure 4.50 Variation of frequency parameter f with
rotating speed Ω ($\alpha = 5°$, $h/a = 0.02$, $L/a = 10$, $n = 2$,
Cone I).

b)　*Influence of cone angle.*

Geometrically, the cone angle is the distinctive difference between conical shells and cylindrical shells. Compared with other geometric parameters, such as the length ratio L/a or thickness ratio h/a, the cone angle has more significant influence on the free vibrations of rotating conical shells. Therefore, in this section, this influence on the frequency characteristics of truncated circular rotating conical shells is examined in detail.

The variations of frequency parameter f against rotating speed Ω for various cone angles α are shown in Fig. 4.19 for a rotating isotropic conical shell. For an orthotropic conical shell (Rem = 3), Fig. 4.20 is generated to investigate the influence of cone angle α on the relationship between frequency parameter f and rotating speed Ω. For the sandwich-type shell defined by Table 4.13, the corresponding results are depicted in

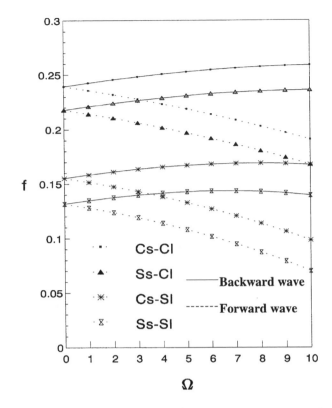

Figure 4.51 Variation of frequency parameter f with
rotating speed Ω ($\alpha = 15°$, $h/a = 0.02$, $L/a = 10$,
$n = 2$, Cone I).

Fig. 4.21. From these figures, it is obvious that the cone angle α has a significant influence
on the frequency parameter f.

For a rotating glass fibre–epoxy composite conical shell with the simply supported
boundary conditions, the influence of cone angle α on frequency characteristics are
shown in Figs. 4.22 and 4.23. Fig. 4.22 presents the influence of cone angle α on the
variation of the frequency parameter f of both backward and forward waves with
circumferential wave number n. From the figure it can be observed that, at small
circumferential wave number n, frequency parameter f decreases with circumferential
wave number n, and this decreasing gradient becomes large with increased values of the
cone angle α. At large circumferential wave number n, the frequency parameter f
increases with circumferential wave number n, and the increasing gradient becomes large
with the decrease in the cone angle α. Fig. 4.23 shows the influence of cone angle α on
the variation of the frequency parameter f with the rotating speed Ω. From the figure it can

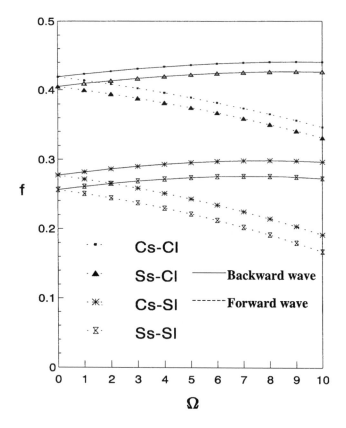

Figure 4.52 Variation of frequency parameter f with rotating
speed Ω ($\alpha = 30°$, $h/a = 0.02$, $L/a = 10$, $n = 2$, Cone I).

be observed that, in the cone angle range $\alpha < 45°$, the frequency parameters f of both backward and forward waves increase with the cone angle α. The rotating composite conical shell with cone angle $\alpha = 45°$ has the highest frequency parameter f for both backward and forward waves. In the cone angle range $\alpha > 45°$, the frequency parameter f decreases with the cone angle α. Moreover, for any cone angle α, the difference in the frequency parameters between the backward and forward waves increases monotonically with increasing rotating speed Ω. The above discussions confirm the significance of the influence of cone angle α on the frequency characteristics in rotating conical shells.

c)   *Influence of length and thickness.*

In this section, we discuss the influence of length ratio L/a and thickness ratio h/a on the frequency characteristics of rotating truncated circular conical shells.

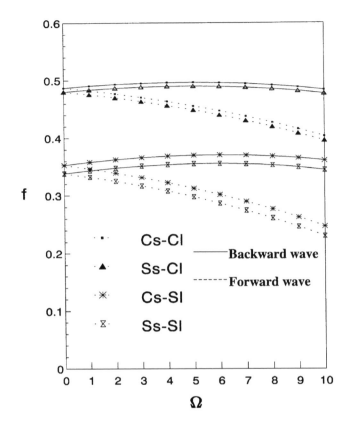

Figure 4.53 Variation of frequency parameter f with rotating speed Ω ($\alpha = 45°$, $h/a = 0.02$, $L/a = 10$, $n = 2$, Cone I).

Figure 4.24 shows the variation of frequency parameter f with the length ratio L/a of the rotating conical shell (Cone III defined in Table 4.14) for four boundary condition cases Cs–Cl, Ss–Cl, Cs–Sl and Ss–Sl, and $\alpha = 30°$, $h/a = 0.005$, $n = 2$ and $\Omega = 4$ rps. From the figure it is obvious that, with increasing L/a values, the frequency parameter f first decreases rapidly, and then in a more gradual manner. The Cs–Cl conical shell has the highest frequency parameter f for both backward and forward waves, followed by the Ss–Cl, Cs–Sl and Ss–Sl shells. The bifurcation of the frequency parameter f at small L/a values is insignificant. The bifurcation becomes increasingly significant as the L/a ratio increases. Moreover, the differences between the frequency parameters for the four boundary conditions are large when the L/a ratio is small. As the L/a ratio increases, the differences between the Cs–Cl and Ss–Cl conical shells, and that between Cs–Sl and Ss–Sl conical shells, become small.

The material orthotropic influence on the relationship between the frequency parameter f and the length ratio L/a is shown in Fig. 4.25 for simply supported rotating

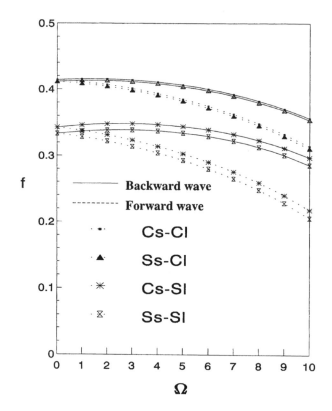

Figure 4.54 Variation of frequency parameter f with
rotating speed Ω ($\alpha = 60°$, $h/a = 0.02$, $L/a = 10$,
$n = 2$, Cone I).

composite conical shells. From the figure it can be seen that the variation of the frequency parameter f with the L/a ratio is steep when orthotropic parameter E^*(Rem) is small. When the orthotropic parameter E^* increases, the steepness of this variation decreases. From the figure it can also be observed that, for the various orthotropic parameters E^*, the frequency parameters f of both backward and forward waves generally decrease with the increase in the length ratio L/a. The difference in the frequency parameters f between the backward and forward waves increases with increasing L/a ratio. When the L/a ratio is small, the decrease in the frequency parameter f with the L/a ratio is rapid. With subsequent increase in the L/a ratio, the decrease in the frequency parameter f with L/a becomes relatively more gradual. Therefore, it can be concluded that the orthotropic influence on the variation of frequency parameter f with the length ratio L/a is significant. If the orthotropic parameter E^* is small, the influence of the L/a ratio on the frequency parameter f is significant for a short conical shell. With increasing length of the conical

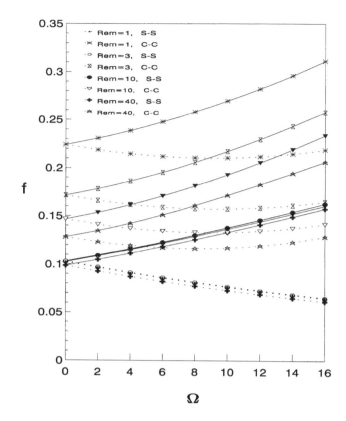

Figure 4.55 Variation of frequency parameter f with rotating velocity Ω (rps) for cone angle $\alpha = 5°$ and different orthotropic parameters Rem ($m = 1$, $n = 1$, $h/a = 0.02$, $L/a = 15$).

shell, the influence of the L/a ratio becomes less significant. For large orthotropic parameter E^*, the influence of material orthotropy on the frequency parameter f is more significant than that of the L/a ratio.

The influence of initial pressure parameter p on the variation of frequency parameter f with the length ratio L/a is shown in Fig. 4.26 for a rotating isotropic conical shell with cone angle $\alpha = 30°$, Cs–Cl boundary condition, rotating speed $\Omega = 5$ rps, and considering circumferential wave number $n = 2$. From the figure it can be seen that, for different magnitudes of initial pressure parameter p, the variation of frequency parameter f of both backward and forward waves with the L/a ratio first decreases rapidly at almost similar gradients for different initial pressure parameters p, after which the frequencies increase gently at different gradients depending on the initial pressure parameter p. The increase in the initial pressure parameter p results in gradient of increasing frequency

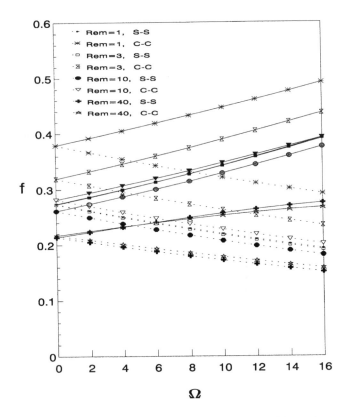

Figure 4.56 Variation of frequency parameter f with rotating
velocity Ω (rps) for cone angle $\alpha = 15°$ and different
orthotropic parameters Rem ($m = 1, n = 1, h/a = 0.02$,
$L/a = 15$).

parameter f being steeper. However, it should also be noted that, when initial pressure parameter $p = 0$, namely, for the rotating conical shell without initial pressure load, the variation of frequency parameter f of both backward and forward waves with the L/a ratio first similarly decreases rapidly and subsequently continues in gradual decrease. The difference in the frequency parameters f between backward and forward waves increases monotonically with the L/a ratio for various initial pressure parameters p. When the L/a ratio is small, the difference between the frequency parameters f for different initial pressure parameters p is small. However, this difference increases rapidly with the increase in the L/a ratio. Therefore, it can be said that the influence of initial uniform pressure load on the variation of frequency parameter f with the length ratio L/a is significant, and this influence becomes more significant for larger L/a ratios. For a given L/a ratio, the increasing magnitude of initial uniform pressure load increases

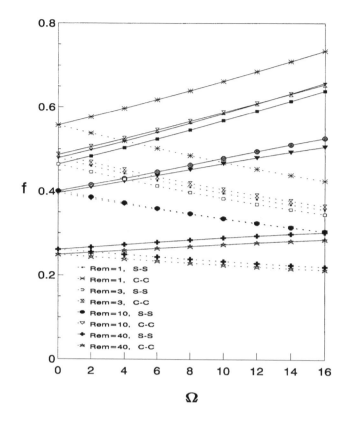

Figure 4.57 Variation of frequency parameter f with rotating velocity Ω (rps) for cone angle $\alpha = 30°$ and different orthotropic parameters Rem ($m = 1$, $n = 1$, $h/a = 0.02$, $L/a = 15$).

the frequency parameters f. For a short rotating conical shell, the influence of the L/a ratio on the frequency characteristics is more significant than that of initial uniform pressure load. However, when the rotating conical shell becomes long, the influence of the magnitude of initial uniform pressure load on the frequency characteristics is much more significant than that of the length ratio L/a.

Figure 4.27 shows the variation of frequency parameter f with the thickness ratio h/a for the four boundary conditions Cs–Cl, Ss–Cl, Cs–Sl and Ss–Sl. Properties of this Cone II shell (defined by Table 4.14) are $\alpha = 15°$, $L/a = 8$, $n = 2$ and $\Omega = 4$ rps. From the figure it can be observed that the Cs–Cl conical shell has the highest frequency parameter f for both backward and forward waves, followed by the Ss–Cl, Cs–Sl and Ss–Sl cases. It is also observed that the increase in the frequency parameter f for both backward and forward waves is very minimal with increasing h/a ratio. Consequently, it can be concluded that

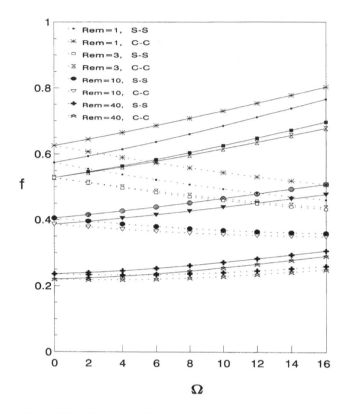

Figure 4.58 Variation of frequency parameter *f* with rotating velocity Ω (rps) for cone angle $\alpha = 45°$ and different orthotropic parameters Rem ($m = 1$, $n = 1$, $h/a = 0.02$, $L/a = 15$).

the influence of the boundary condition on frequency characteristics is much more significant than that of the h/a ratio. Effectively, the influence of the h/a ratio of thin rotating conical shells on frequency characteristics may even be negligible.

The material orthotropic influence on the relationship between the frequency parameter *f* and thickness ratio h/a is shown in Fig. 4.28 for a simply supported rotating composite conical shell. From the figure it can be seen that the increase in the frequency parameter *f* of both backward and forward waves is very minimal with increasing h/a ratio. The material orthotropic influence on the variation of frequency parameter *f* with the h/a ratio can also be seen. Therefore, it can be said that the orthotropic influence on frequency characteristics is much more significant than that of the thickness ratio h/a, where the effects are almost negligible.

The influence of initial pressure parameter *p* on the relationship between the frequency parameter *f* and the thickness ratio h/a is shown in Fig. 4.29 for a clamped

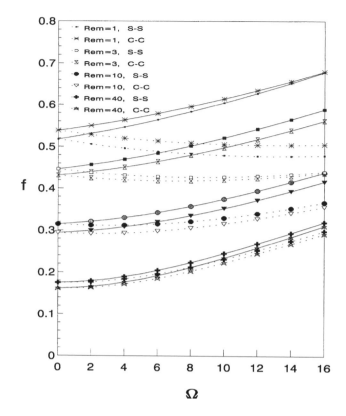

Figure 4.59 Variation of frequency parameter f with rotating
velocity Ω (rps) for cone angle $\alpha = 60°$ and different
orthotropic parameters Rem ($m = 1$, $n = 1$, $h/a = 0.02$,
$L/a = 15$).

rotating isotropic conical shell having cone angle $\alpha = 45°$ and rotating speed $\Omega = 10$ rps, with the consideration of circumferential wave number $n = 2$. From the figure it can be observed that, for a rotating conical shell with no initial pressure load, the increase in the frequency parameter f of both backward and forward waves is monotonic and very minimal with increasing h/a ratio. However, for a rotating conical shell with initial uniform pressure load, the variation of frequency parameter f of both backward and forward waves with the h/a ratio first decreases rapidly but subsequently becomes gradual. The difference in the frequency parameters f for different initial pressure parameters p is large for small h/a ratios. With the increase in the h/a ratio, these differences decrease. Therefore, it can be concluded that the influence of initial uniform pressure load on the relationship between the frequency parameter f and the thickness ratio h/a is significant. Such an influence becomes more significant at small h/a ratios. For a rotating conical shell

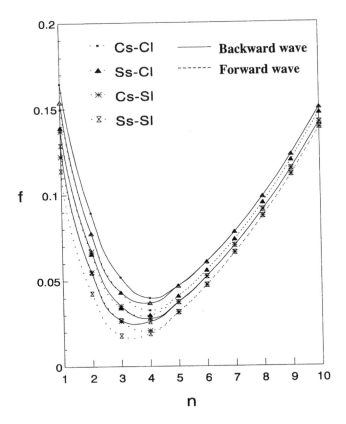

Figure 4.60 Variation of frequency parameter f with
circumferential wave number n ($\alpha = 5°$, $h/a = 0.015$,
$L/a = 10$, $\Omega = 3$ rps, Rem $= 3$).

without initial pressure load, the variation of frequency parameter f with the h/a ratio shows general minimal increase in a monotonic manner. However, for a rotating conical shell with initial uniform pressure load, the frequency parameter f first decreases rapidly which subsequently becomes very gradual. Generally, increasing the magnitude of the initial uniform pressure load increases the frequency parameter f. Also the influence of initial uniform pressure load on the frequency characteristics is more significant than that of the thickness ratio h/a for a rotating conical shell.

d)   ***Influence of orthotropy and layered configuration of composites.***

In the analysis of laminated composite shells, material orthotropy is usually an important physical parameter. A study is thus conducted here to examine the influence

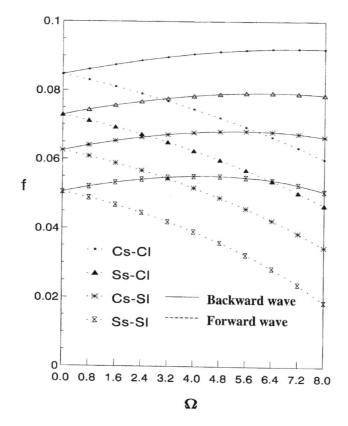

Figure 4.61 Variation of frequency parameter f with rotating speed Ω ($\alpha = 5°$, $h/a = 0.015$, $L/a = 10$, $n = 2$, Rem $= 3$).

of the orthotropic parameter E^* ($=$ Rem), defined by Eq. (4.58), on the frequency characteristics of rotating thin truncated circular laminated composite conical shell. Investigations include the orthotropic E^* influence on the relationship between the frequency parameter f and the circumferential wave number n, and that between the frequency parameter f and the rotating speed Ω. Additionally, the effect of layered configuration of the laminated conical shell on the frequency parameter f is also studied. For simplification of the present discussion, only regular symmetrical cross-ply laminated composite conical shells are considered. Based on the theory of laminated composites, we know that there is a physical equivalence of the constitutive relationship between a single orthotropic layer and regular symmetrical cross-ply laminates. Therefore, in the following discussion for the influence of orthotropic property on frequency characteristics, we shall treat the regular symmetrical cross-ply laminated composite conical shell as an equivalent orthotropic single-layer conical shell.

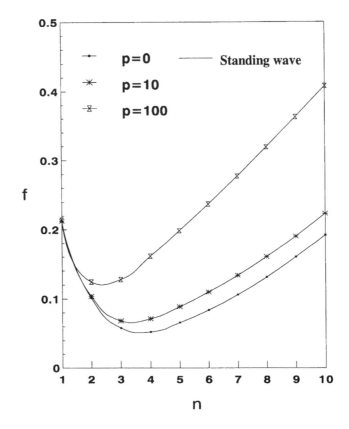

Figure 4.62 Variation of frequency parameter f with circumferential wave number n for a stationary conical shell with Cs–Cl boundary condition ($\alpha = 5°$, $h/a = 0.01$, $L/a = 10$, $\Omega = 0$).

For a stationary laminated composite conical shell, Fig. 4.30 shows the variation of the frequency parameter f of a standing wave motion with circumferential wave number n for four different orthotropic parameters namely, $E^* = 1$, 3, 10 and 40. With increasing circumferential wave number n, the frequency parameter f of a standing wave first decreases and then increases monotonically. At small circumferential wave number n, the gradients of decreasing frequency parameters f of standing waves are almost similar for the different orthotropic parameters E^*. At large circumferential wave number n, the gradients of the increasing frequency parameters f of the standing waves become smaller with increasing orthotropic parameter E^*. Thus, the orthotropic influence on the frequency characteristics of standing waves is significant when circumferential wave number n is large. When considering low circumferential wave number n,

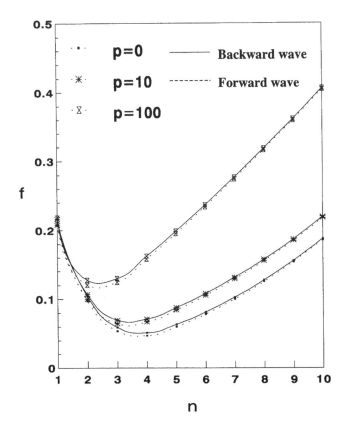

Figure 4.63 Variation of frequency parameter f with
circumferential wave number n for a rotating conical shell
with Cs–Cl boundary condition ($\alpha = 5°$, $h/a = 0.01$,
$L/a = 10$, $\Omega = 2$ rps).

the material orthotropic influence becomes insignificant. The increase in the orthotropy
results in frequency parameter f decrease for the standing wave motion.

In the rotating laminated composite conical shell, for various orthotropic
parameters E^* and at different rotating speeds Ω, the variations of the frequency
parameters f of both backward and forward travelling waves with circumferential wave
number n are shown in Fig. 4.31 ($\Omega = 0.5$ rps) and Fig. 4.32 ($\Omega = 1$ rps). Obviously, the
present variation of the frequency parameters f of both backward and forward waves for
the rotating shell is similar to that of the corresponding standing waves in Fig. 4.30. By
comparing the two later figures, it can be seen that, at small circumferential wave number
n, for different orthotropic parameters E^* and rotating speeds Ω, the frequency
parameters f decrease at almost similar gradients. At large circumferential wave number

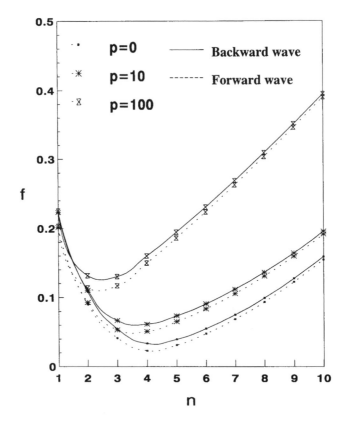

Figure 4.64 Variation of frequency parameter f with
circumferential wave number n for a rotating conical shell
with Cs–Cl boundary condition ($\alpha = 5°$, $h/a = 0.01$,
$L/a = 10$, $\Omega = 5$ rps).

n, the smaller the orthotropic parameter E^*, the more rapid the increase in the frequency parameter f will be. Also, the larger the orthotropic parameter E^*, the more significant is the influence of rotating speed Ω on the frequency characteristics. Therefore, it can be concluded that the influences of orthotropic property and rotating speed on the frequency characteristics of both backward and forward travelling waves are insignificant at small circumferential wave number n. When circumferential wave number n is large, the orthotropic influence becomes very significant. With increasing orthotropic parameter E^*, the increase in the frequency parameter f becomes rather gradual, and the influence of rotating speed on the frequency characteristics thus becomes larger. It is also noted that the frequency parameters f of both backward and forward waves generally decrease with increasing orthotropic parameter E^*.

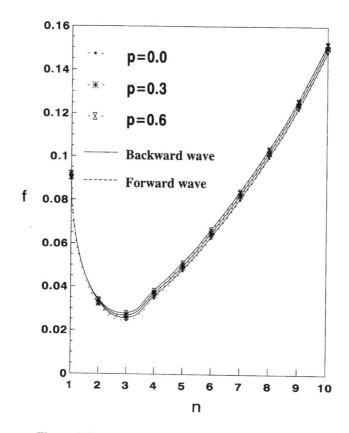

Figure 4.65 Variation of frequency parameter f with
circumferential wave number n for a rotating conical shell
with Ss–Sl boundary condition ($\alpha = 5°$, $h/a = 0.01$,
$L/a = 15$, $\Omega = 0.4$ rps).

For various cone angles α and orthotropic parameters E^*, the relationships between
frequency parameter f and circumferential wave number n are shown in Figs. 4.33–4.37,
which correspond, respectively, to the cases of $\alpha = 5, 15, 30, 45$ and $60°$. By comparing
these figures for the case where rotating speed $\Omega = 0.2$ rps, it can be seen that, at small
circumferential wave number n, the frequency parameter f of both backward and forward
waves decreases and the difference of frequency parameters f for different orthotropic
parameters E^* is small for small cone angles α. With the increase in the cone angle α, the
frequency parameter f of both backward and forward waves initially shows gradual
changes and then decreases rapidly. Also, at low circumferential wave number n and high
cone angles α, difference in the frequency parameters f for different orthotropic
parameters E^* become large. At large circumferential wave number n, the frequency
parameters f of both backward and forward waves increase monotonically and changes in

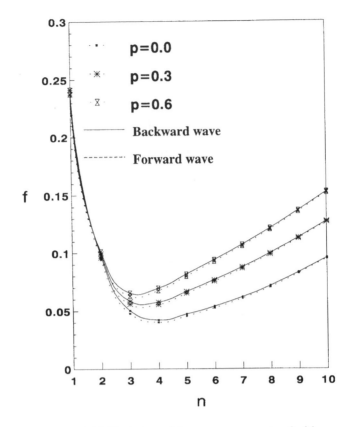

Figure 4.66 Variation of frequency parameter f with
circumferential wave number n for a rotating conical shell
with Ss–Sl boundary condition ($\alpha = 15°$, $h/a = 0.01$,
$L/a = 15$, $\Omega = 0.4$ rps).

these frequency parameters f for different orthotropic parameters E^* are very significant
when cone angles α are small. With the increase in the cone angle α, the variation of
frequency parameter f of both backward and forward waves with circumferential wave
number n (at high n values) changes from positive gradients to negative gradients, and
differences between frequency parameters f for different orthotropic parameters E^*
becoming smaller. Therefore, it can be said that at large circumferential wave number n,
the orthotropic influence on the frequency characteristics of the rotating composite
conical shell with small cone angle α is significant. If both the cone angle α and
circumferential wave number n are small, such influence becomes insignificant. With
large cone angles α, the orthotropic influence on the frequency characteristics at small
circumferential wave number n is much more significant than that at large circumferential
wave number n. In any case, the frequency parameter f of both backward and forward

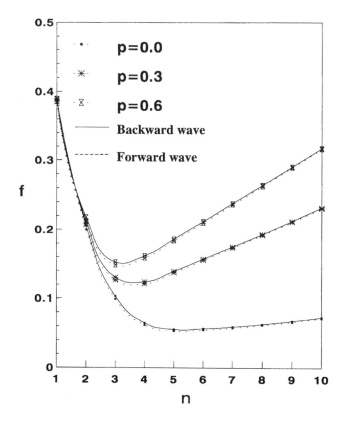

Figure 4.67 Variation of frequency parameter f with
circumferential wave number n for a rotating conical shell
with Ss–Sl boundary condition ($\alpha = 30°$, $h/a = 0.01$,
$L/a = 15$, $\Omega = 0.4$ rps).

waves for small orthotropic parameter E^* is always larger than that for large orthotropic parameter E^*.

For various cone angles α and orthotropic parameters E^*, the relationships between the frequency parameter f and rotating speed Ω are shown in Figs. 4.38–4.42, which correspond, respectively, to the cases of $\alpha = 5$, 15, 30, 45 and 60°. By comparing these figures for the case where circumferential wave number $n = 2$, it can be seen that, for small cone angle α, the material orthotropic influence on the frequency characteristics is insignificant. The differences in frequency parameters f between backward and forward waves increase monotonically with the increase in the rotating speed Ω, for the different orthotropic parameters E^*. Also, the frequency parameters f of the backward waves increase monotonically while those of the forward waves decrease monotonically. When

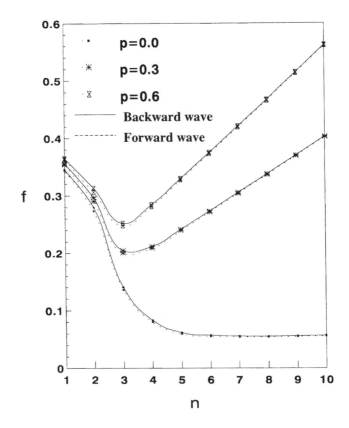

Figure 4.68 Variation of frequency parameter f with
circumferential wave number n for a rotating conical shell
with Ss–Sl boundary condition ($\alpha = 45°$, $h/a = 0.01$,
$L/a = 15$, $\Omega = 0.4$ rps).

cone angle α becomes large, the orthotropic influence on frequency characteristics becomes very significant, and in addition the differences in frequency parameters f between backward and forward waves decrease with the increase in the orthotropic parameter E^*. Therefore, it can be said that the material orthotropic influence on the relationship between the frequency parameter f and rotating speed Ω is significant at large cone angle α. The smaller the cone angle α, the less significant such orthotropic influence is. It should also be noted that when the cone angle α is large, the increase in the orthotropic parameter E^* results in the decrease in the difference between the frequency parameters f of the backward and forward waves.

Based on the above studies, it can be concluded that the material orthotropic property has substantial influence on the frequency characteristics of rotating composite

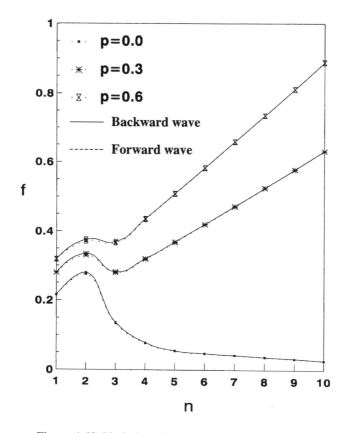

Figure 4.69 Variation of frequency parameter f with
circumferential wave number n for a rotating conical shell
with Ss–Sl boundary condition ($\alpha = 60°$, $h/a = 0.01$,
$L/a = 15$, $\Omega = 0.4$ rps).

conical shells. Such orthotropic influence is significant in cases of large cone angle or large circumferential wave number. The increase in the orthotropic property results in the frequency parameter decrease for both the backward and forward waves.

Here we investigate the influence of layered configuration in various sandwich-type rotating conical shells on the frequency characteristics. Let us consider three conical shells with different layered configurations, as shown in Table 4.14. The three sandwich-type conical shells, which have three layers of construction are termed Cones I, II and III to differentiate their different layered configurations. Cone I has a middle layer that is three times the thickness of the two surface layers; Cone II has a thin middle layer that is three times thinner than the two surface layers; Cone III has three layers each of similar thickness. Their respective thicknesses and material properties are provided in Table 4.14.

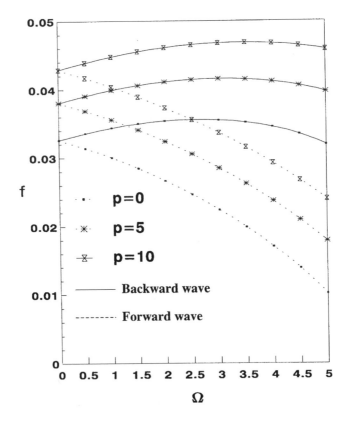

Figure 4.70 Variation of frequency parameter *f* with rotating
speed Ω for a rotating conical shell with Ss–Sl boundary
condition ($\alpha = 5°$, $h/a = 0.01$, $L/a = 15$, $n = 2$).

To investigate the influence of the layered configuration in the three sandwich-type
conical shells on the frequency characteristics, we will consider the case where $\alpha = 5°$
and for four boundary conditions, namely Cs–Cl, Ss–Cl, Cs–Sl and Ss–Sl. The
relationships between frequency parameter *f* and rotating speed Ω are shown in Fig. 4.43
(Cone I), Fig. 4.44 (Cone II) and Fig. 4.45 (Cone III). Comparing these figures, it can be
seen that the influence of the layered configuration of the thin sandwich-type conical
shells on the frequency parameter of the forward wave is slightly larger than that of the
backward wave by increasing the rotating speed Ω. Although the influence of the layered
configuration on the variation of frequency parameter *f* with the rotating speed Ω is
insignificant, it is observed that Cone I, which has a thicker middle layer than the surface
layers, has the relative highest frequency parameter *f* of both backward and forward
waves, followed by Cone III, which has layers of equal thickness, and Cone II, which has

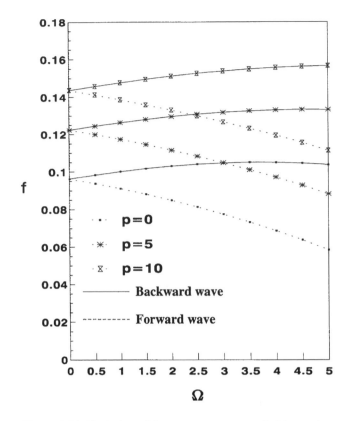

Figure 4.71 Variation of frequency parameter f with rotating speed Ω for a rotating conical shell with Ss–Sl boundary condition ($\alpha = 15°$, $h/a = 0.01$, $L/a = 15$, $n = 2$).

a thinner middle layer than the two surface layers. Therefore, when considering the sandwich-type conical shells, it can be said that, for small cone angle, the influence of boundary condition on the frequency characteristics is much more significant than that of the layered configuration.

e)   Influence of boundary condition.

In the analysis of shell structures, it is common for the bending boundary conditions to receive the most attention, and two of the more common edge conditions are the clamped and simply supported boundary conditions. They are considered herein in all cases. For both of these boundary conditions, the out-of-shell displacement normal to the reference surface is zero. In addition, the out-of-shell rotation of a material line element normal to a clamped edge is also zero. Further, for a simply supported edge,

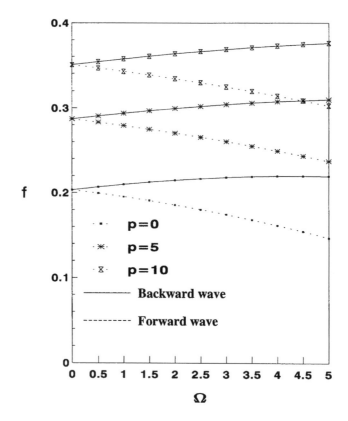

Figure 4.72 Variation of frequency parameter *f* with rotating
speed Ω for a rotating conical shell with Ss–Sl boundary
condition ($\alpha = 30°$, $h/a = 0.01$, $L/a = 15$, $n = 2$).

the component of the bending moment vector, which is tangent to the shell edge, is zero, such that the shell support is analogous to a frictionless hinge. It should be noted here that there are various forms of the simply supported boundary condition due to different physical and engineering considerations.

However, in all cases herein for rotating conical shells, we will consider four different combinations of boundary conditions. They are the simply supported small edge–clamped large edge (Ss–Cl), clamped small edge–simply supported large edge (Cs–Sl), both edges clamped (Cs–Cl or C–C), and both edges simply supported (Ss–Sl or S–S) boundary conditions.

For the discussion on rotating sandwich-type conical shell (Cone I defined in Table 4.14), two studies are carried out on the effect of boundary conditions on frequency parameter *f*. These are the effect on the relationship between frequency parameter *f* and circumferential wave number *n*, as shown in Figs. 4.46–4.49, and the effect on

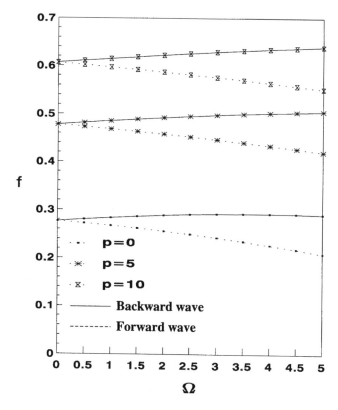

Figure 4.73 Variation of frequency parameter f with rotating speed Ω for a rotating conical shell with Ss–Sl boundary condition ($\alpha = 45°$, $h/a = 0.01$, $L/a = 15$, $n = 2$).

the variation of the frequency parameter f with the rotating speed Ω, as shown in Figs. 4.50–4.54.

Figure 4.46 shows that, for all the four boundary condition cases considered, when the present sandwich-type conical shell is stationary, the frequency parameter f of a standing wave motion first decreases rapidly, and then increases monotonically with increasing circumferential wave number n. The Cs–Cl conical shell has the highest frequency parameter f, followed by the Ss–Cl, Cs–Sl and Ss–Sl shells. At small circumferential wave numbers, we observe a relatively large difference between the frequency parameters of the four boundary conditions, implying that the influence of the boundary condition is significant. However, the difference in frequency parameters between the Cs–Cl and Ss–Cl shells or between the Cs–Sl and Ss–Sl shells diminishes with increasing circumferential wave number n. For large circumferential wave number n, the Cs–Cl and Ss–Cl boundary condition cases converge and so do the Cs–Sl and

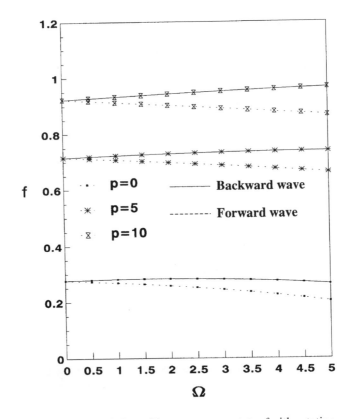

Figure 4.74 Variation of frequency parameter f with rotating speed Ω for a rotating conical shell with Ss–Sl boundary condition ($\alpha = 60°$, $h/a = 0.01$, $L/a = 15$, $n = 2$).

Ss–Sl boundary condition cases. However, the frequency parameters f of the two former boundary condition cases are always larger than those of the latter two boundary condition cases.

When the conical shell with cone angle $\alpha = 5°$ rotates at different rotating speeds and for various geometric properties, Figs. 4.47–4.49 show that, for the four boundary conditions considered, the variation of the frequency parameters f of both backward and forward travelling waves with circumferential wave number n is similar to that of the non-rotating conical shell as shown in Fig. 4.46. The rotating Cs–Cl conical shell also has the highest frequency parameter f, followed similarly by the corresponding Ss–Cl, Cs–Sl and Ss–Sl shells. Figure 4.47 is for the case of rotating speed $\Omega = 3$ rps and geometric properties of $h/a = 0.01$ and $L/a = 8$; Fig. 4.48 for $\Omega = 4$ rps, $h/a = 0.015$ and $L/a = 10$; and Fig. 4.49 for $\Omega = 5$ rps, $h/a = 0.02$ and $L/a = 10$. Comparing these three figures for different rotating speeds, it can be seen that, with increasing

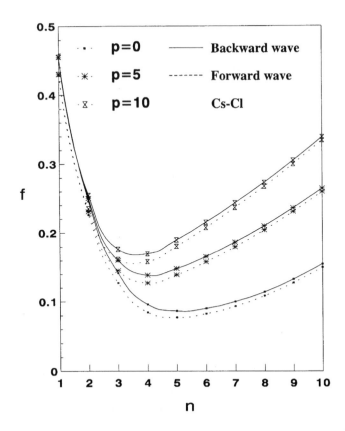

Figure 4.75 Variation of frequency parameter f with
circumferential wave number n for a rotating conical shell
with Cs–Cl boundary condition ($\alpha = 15°$, $h/a = 0.015$,
$L/a = 10$, $\Omega = 3$ rps).

circumferential wave number n, the variations of the frequency parameters f of the
backward and forward waves first decrease at similar gradients, then increase at gradients
which increase with the rotating speed. It can also be observed that the differences
between the frequency parameters of the four boundary condition cases are relatively
higher at small circumferential wave number. With the increase in the circumferential
wave number n, the frequency parameters of Cs–Cl and Ss–Cl conical shells for both
backward and forward waves converge, as do those of the Cs–Sl and Ss–Sl conical
shells. Consequently, it can be concluded that the influence of boundary condition on a
rotating conical shell is significant only at a few small circumferential wave numbers n.
When circumferential wave number n increases, the influence of rotating speed Ω on the
relationship between the frequency parameter f and circumferential wave number n
becomes more significant than that of boundary condition.

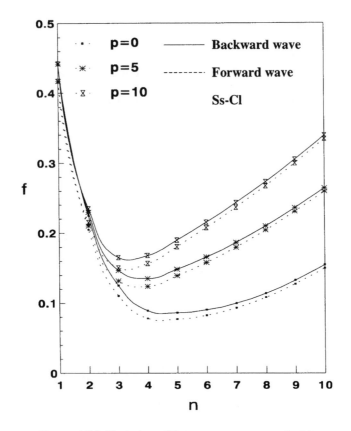

Figure 4.76 Variation of frequency parameter f with
circumferential wave number n for a rotating conical shell
with Ss–Cl boundary condition ($\alpha = 15°$, $h/a = 0.015$,
$L/a = 10$, $\Omega = 3$ rps).

To examine the influence of boundary condition on the relationship between the frequency parameter f and the rotating speed Ω, the frequency characteristics for five different cone angles ($\alpha = 5$, 15, 30, 45 and 60°) are shown in Figs. 4.50–4.54, respectively. Comparing these figures, it can be seen that, when the cone angle is small, the frequency parameters f of the backward waves increase while those of forward waves decrease, with the increase in the rotating speed Ω. Also, the differences in the frequency parameters f between backward and forward waves increase monotonically. For large cone angles, the frequency parameters f of both backward and forward waves decrease with the increase in the rotating speed Ω. In addition, with the increase in the cone angle, the differences in the frequency parameters between the Cs–Cl and Ss–Cl conical shells, and those between the Cs–Sl and Ss–Sl conical shells, diminishes. Generally, the frequency parameters f associated

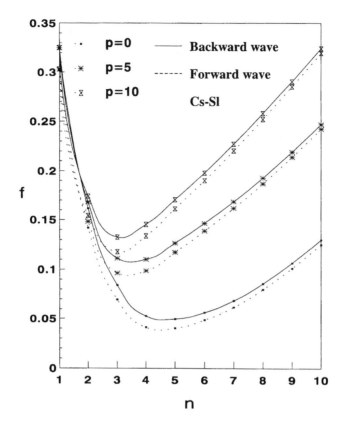

Figure 4.77 Variation of frequency parameter f with circumferential wave number n for a rotating conical shell with Cs–Sl boundary condition ($\alpha = 15°$, $h/a = 0.015$, $L/a = 10$, $\Omega = 3$ rps).

with the first two boundary conditions are higher than those associated with the latter two boundary conditions. Also, the rotating Cs–Cl conical shell possesses the highest frequency parameter f, followed by the Ss–Cl, Cs–Sl and Ss–Sl shells. It can thus be concluded that, for a rotating conical shell with small cone angle, the influence of the boundary condition on the variation of frequency parameter f with rotating speed Ω is significant. When the cone angle becomes large, the influences of the Cs–Cl and Ss–Cl boundary condition cases on the frequencies are equivalent. Similar observations are made for the Cs–Sl and Ss–Sl shells.

    To investigate the coupling influence of material orthotropy and boundary condition on the free vibration of rotating orthotropic conical shells, the frequency characteristics for five different cone angles α are shown in Figs. 4.55–4.59. As shown in the figures, the variations of frequency parameter f with rotating speed Ω are presented

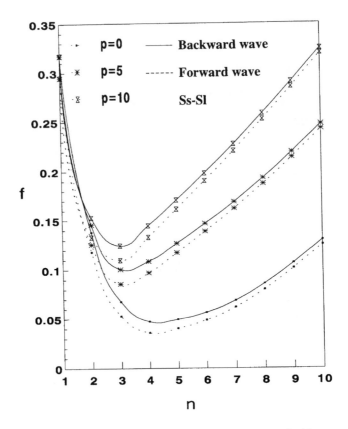

Figure 4.78 Variation of frequency parameter f with
circumferential wave number n for a rotating conical shell
with Ss–Sl boundary condition ($\alpha = 15°$, $h/a = 0.015$,
$L/a = 10$, $\Omega = 3$ rps).

for different orthotropic parameters E^* (= Rem) and for both S–S (or Ss–Sl) and C–C (or Cs–Cl) boundary conditions. These five figures correspond to the five different cone angles, namely, $\alpha = 5$, 15, 30, 45 and 60°. Comparing Figs. 4.55–4.59, it can be seen that, for small cone angle α, the influence of orthotropic parameter E^* on the frequency characteristics for the C–C boundary conditions is more significant than that for the S–S boundary conditions. When cone angle α is large, the influence of orthotropic parameter E^* on the frequency characteristics is more significant than that of boundary condition. Moreover, for small cone angle α, with increasing rotating speed Ω, the frequency parameters f of the backward waves increase and those of the forward waves decrease. However, for large cone angle α, the same can only be said for cases involving small orthotropic parameter E^*. With increasing orthotropic parameter E^*, the frequency parameters f of both backward and forward waves increase. For the case of a rotating

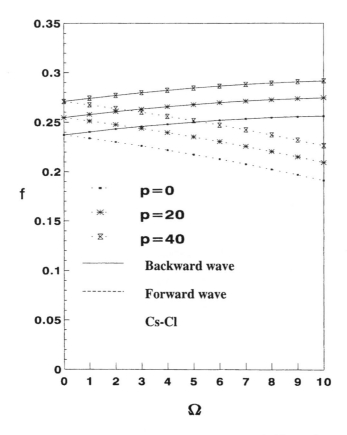

Figure 4.79 Variation of frequency parameter f with rotating
speed Ω for a rotating conical shell with Cs–Cl boundary
condition ($\alpha = 15°$, $h/a = 0.015$, $L/a = 10$, $n = 2$).

glass fibre–epoxy laminated composite conical shell with cone angle $\alpha = 5°$,
the influences of boundary condition on the frequency characteristics are shown in
Figs. 4.60 and 4.61. Figure 4.60 shows the influence of boundary condition on
the variation of frequency parameter f with circumferential wave number n while Fig. 4.61
shows the influence of boundary condition on the variation of frequency parameter f with
rotating speed Ω. Similar phenomena and conclusions as those corresponding to
Figs. 4.46–4.50 can be observed and drawn, respectively.

f)   ***Influence of initial stress.***

In the engineering design of industrial boilers and pressure vessels, the shell
structures are subjected to uniform pressure load, such as an internal or external

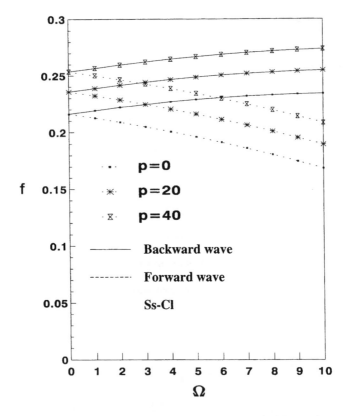

Figure 4.80 Variation of frequency parameter *f* with rotating
speed Ω for a rotating conical shell with Ss–Cl boundary
condition ($\alpha = 15°$, $h/a = 0.015$, $L/a = 10$, $n = 2$).

hydrostatic or gas pressure. When the conical shell structures are stationary and subjected
to the uniform pressure load, unlike corresponding cylindrical shell, the stress field
generated in the conical shell varies with the spatial coordinates. The stress levels in the
conical shell structure are thus of critical importance to designers.

When we consider the free vibration of a rotating thin truncated circular isotropic
conical shell under an initial uniform pressure load, the stresses in the shell consist of the
initial stress arising from the initial pressure load and additional vibrational stresses such
as those due to the rotation. In order to simplify the present formulation and uncouple the
initial and vibrational stresses, it is assumed that the bending stress in the initial loading
state and the displacements due to the membrane stress are very small and are thus
neglected, and the pressure direction also remains in its initial direction during vibration.
Additionally, in assuming the initial stress field being in equilibrium, the state of static

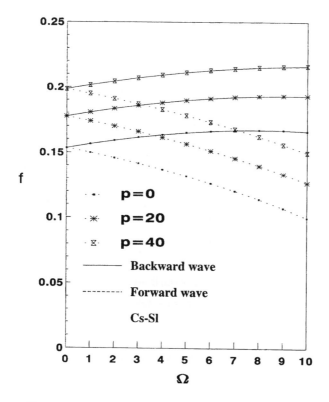

Figure 4.81 Variation of frequency parameter f with
rotating speed Ω for a rotating conical shell with Cs–Sl
boundary condition ($\alpha = 15°$, $h/a = 0.015$, $L/a = 10$,
$n = 2$).

equilibrium may, therefore, be taken as a reference state for the dynamic deformation due
to vibration.

By the GDQ method, the influence of initial uniform pressure on the frequency
characteristics of a rotating conical shell is studied. A non-dimensional initial pressure
parameter p, defined by Eq. (4.59), is employed for the present discussion. Four studies
on the frequency characteristics are carried out here. The first is the study of the
influences of initial pressure parameter p for different rotating speeds Ω on the
relationship between frequency parameter f and circumferential wave number n in
rotating conical shells with cone angle $\alpha = 5°$ and Cs–Cl boundary condition. The results
are shown in Figs. 4.62–4.64. The second is on the influence of initial pressure parameter
p on the relationship between frequency parameter f and circumferential wave number n
in rotating Ss–Sl conical shells for five different cone angles, $\alpha = 5, 15, 30, 45$ and $60°$.

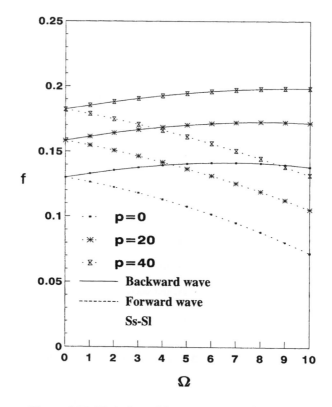

Figure 4.82 Variation of frequency parameter f with rotating speed Ω for a rotating conical shell with Ss–Sl boundary condition ($\alpha = 15°$, $h/a = 0.015$, $L/a = 10$, $n = 2$).

The results are shown in Figs. 4.65–4.69. The third is on the influence of initial pressure parameter p on the relationship between frequency parameter f and rotating speed Ω in rotating Ss–Sl conical shell for different cone angles of $\alpha = 5$, 15, 30, 45 and 60°, respectively. The results are shown in Figs. 4.70–4.74. The fourth is the coupling influence of the initial pressure parameter p and the boundary condition on the frequency characteristics of rotating conical shells with cone angle $\alpha = 15°$. The results are shown in Figs. 4.75–4.82.

For a stationary conical shell with the cone angle $\alpha = 5°$ and Cs–Cl boundary condition, Fig. 4.62 presents the variation of the frequency parameter f of the standing wave motion with circumferential wave number n for three different non-dimensional initial pressure parameters, $p = 0$, 10 and 100. From the figure it can be seen that, with increasing circumferential wave number n, the frequency parameters f of the standing wave first

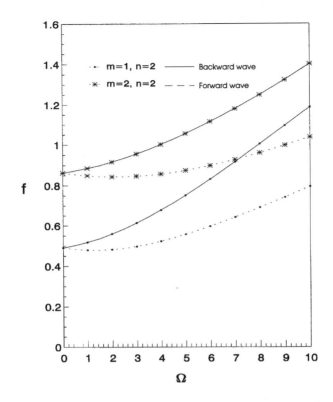

Figure 4.83 Relationship between frequency parameter f
and rotating speed Ω for various modes of free vibration
$(\mu = 0.3, \alpha = 45°, h/a = 0.02, L/a = 20)$.

decrease and then increase monotonically for a given initial pressure parameter p. The gradient of the increasing frequency parameters f becomes steeper with increasing initial pressure parameter p. Meanwhile, with the increase in the initial pressure parameter p, frequency parameters f also increase in a monotonic manner, for any given circumferential wave number n. Therefore, it can be concluded that, for a non-rotating conical shell, the influence of initial uniform pressure load on the standing wave frequency characteristics is more significant when circumferential wave number n is large.

When the conical shell considered in Fig. 4.62 starts to rotate, the variations of the frequency parameters f of both backward and forward travelling waves with circumferential wave number n for different initial pressure parameters p are shown in Fig. 4.63 ($\Omega = 2$ rps) and Fig. 4.64 ($\Omega = 5$ rps). Comparing these three figures, it is evident that the variation of the frequency parameters f of both backward and forward waves for the rotating conical shell is very similar to that of the standing wave

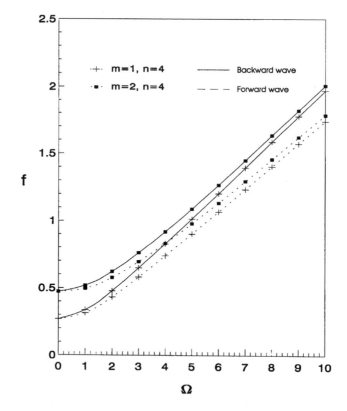

Figure 4.84 Relationship between frequency parameter f
and rotating speed Ω for various modes of free vibration
($\mu = 0.3$, $\alpha = 45°$, $h/a = 0.02$, $L/a = 20$).

of the non-rotating shell. It can also be seen that, with the increase in the rotating speed Ω, the differences in frequency parameters f between the backward and forward waves increase. For large circumferential wave number n, if initial pressure parameter p is small, the influence of rotating speed Ω on the relationship between frequency parameter f and circumferential wave number n is significant. When the initial pressure parameter p becomes large, the influence of initial pressure parameter p on the frequency parameter f becomes more significant than that of the rotating speed Ω. For a given rotating speed Ω, the frequency parameters f of both backward and forward waves decrease at small circumferential wave number n, and the gradients of these decreases are quite similar for different initial pressure parameters p. At large circumferential wave number n, the frequency parameters f increase monotonically, and the gradients of these increases become steeper with increasing initial pressure parameter p. We also observe that the frequency parameter f generally increases with increasing initial uniform pressure load.

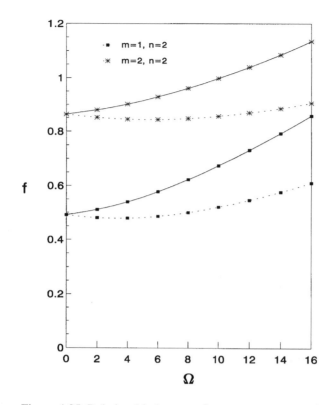

Figure 4.85 Relationship between frequency parameter f
and rotating speed Ω (rps, revolutions per second or Hz)
at various vibrational modes ($\alpha = 45°$, $h/a = 0.008$,
$L/a = 20$).

Therefore, it can be concluded that the influences of initial uniform pressure load and rotating speed on the frequency characteristics of both backward and forward travelling waves are significant at large circumferential wave number n. However, with the increase in the initial uniform pressure load, its influence becomes more significant than that of the rotating speed Ω.

For the rotating Ss–Sl conical shell under different initial pressure parameters p, the relationship between frequency parameter f and circumferential wave number n are shown in Figs. 4.65–4.69 for cone angle cases of $\alpha = 5$, 15, 30, 45 and 60°, respectively. Comparing these figures for the case of rotating speed $\Omega = 0.4$ rps, it is observed that, for small circumferential wave number n and cone angle α, the frequency parameter f of both backward and forward waves decreases monotonically, and the difference in the frequency parameters f for different initial pressure parameters p is small. When the cone angle α becomes large, the frequency parameters f no longer decrease at low circumferential wave

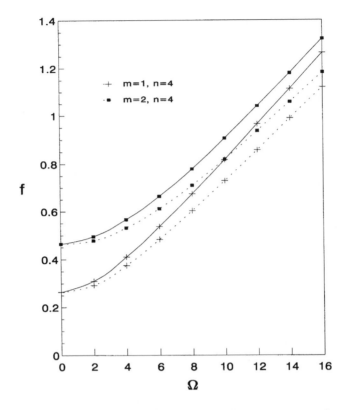

Figure 4.86 Relationship between frequency parameter f
and rotating speed Ω (rps, revolutions per second or Hz)
at various vibrational modes ($\alpha = 45°$, $h/a = 0.008$,
$L/a = 20$).

number n, and the difference in frequency parameters f for different initial pressure
parameters p becomes large. However, for large circumferential wave number n, when cone
angle α is small, the frequency parameters f of both backward and forward waves increase
monotonically. With the increase in the cone angle α, the difference of frequency
parameters f for different initial pressure parameters p becomes progressively larger. When
cone angle α is large, the influence of initial pressure parameter p on the frequency
characteristics is very significant. Therefore, it can be concluded that, for large cone angle α,
the influence of initial uniform pressure load on the relationship between frequency
parameter f and circumferential wave number n is much more significant than that for small
cone angle α. For a given cone angle α, the influence of initial uniform pressure load for
large circumferential wave number n is more significant than that for small circumferential
wave number n. For a given circumferential wave number n, an increase in the magnitude

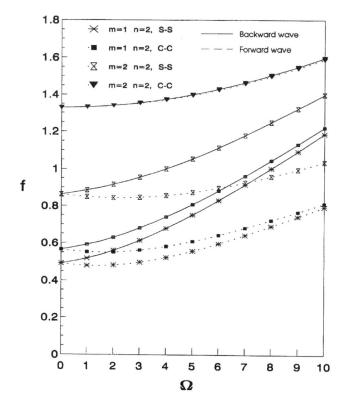

Figure 4.87 Variation of the frequency parameter f with the
rotating velocity Ω (rps) at various vibrational modes
($\mu = 0.3$, $\alpha = 45°$, $h/a = 0.02$, $L/a = 20$).

of initial uniform pressure load will result in an increase in the frequency parameter f. At large circumferential wave number n, higher magnitudes of the initial pressure load are associated with steeper gradients of increase in the frequency parameter f.

For the rotating Ss–Sl conical shell under different initial pressure parameters p, the relationship between the frequency parameter f and the rotating speed Ω is shown in Figs. 4.70–4.74 for cone angle cases of $\alpha = 5$, 15, 30, 45 and 60°, respectively. By comparing these figures for the case of circumferential wave number $n = 2$ it can be observed that, with the increase in the cone angle α, the differences in frequency parameters f increase monotonically with the initial pressure parameters p. For a given cone angle α, with the increase in the rotating speed Ω, the frequency parameter f of the backward wave increases and that of the forward wave decreases so that the differences in frequency parameters f between the backward and forward waves increase monotonically. Such an increase in the frequency difference with the rotating

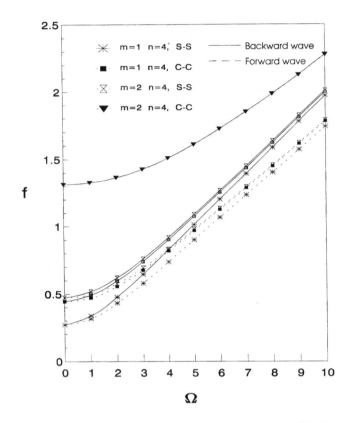

Figure 4.88 Variation of the frequency parameter f with the rotating velocity Ω (rps) at various vibrational modes ($\mu = 0.3$, $\alpha = 45°$, $h/a = 0.02$, $L/a = 20$).

speed Ω is relatively more rapid for small cone angle α, and it becomes gentler with the increase in the cone angle α. Therefore, it can be said that the influence of initial uniform pressure load for a large cone angle α on the relationships between frequency parameter f and rotating speed Ω is more significant than that for a small cone angle α. For a given rotating speed Ω, the difference in frequency parameters f for different magnitudes of initial uniform pressure load increases with increasing cone angle α. Generally, an increase in the magnitude of the initial uniform pressure load will result in an increase in the frequency parameters f.

For the discussion on coupled influence of initial pressure parameter p and boundary condition on the frequency characteristics, Figs. 4.75–4.82 are generated for a rotating thin truncated circular isotropic conical shell with the cone angle $\alpha = 15°$. In these figures, the boundary conditions considered are the Cs–Cl, Ss–Cl, Cs–Sl and Ss–Sl. The present discussions of the figures are divided into two parts. The first for rotating speed set

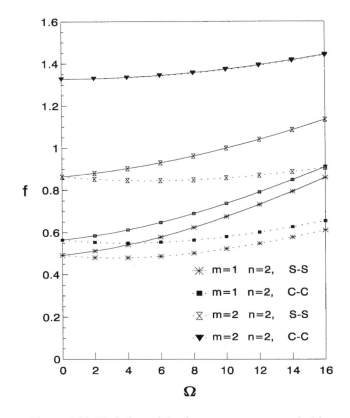

Figure 4.89 Variation of the frequency parameter f with the rotating velocity Ω (rps) at different vibrational modes for a sandwich-type conical shell ($\alpha = 45°$, $h/a = 0.008$, $L/a = 20$).

at $\Omega = 3$ rps, as shown in Figs. 4.75–4.78, presents the relationship between frequency parameter f and circumferential wave number n, for the four different boundary conditions and three initial pressure parameters p. The second with circumferential wave number set at $n = 2$, as shown in Figs. 4.79–4.82, presents the variation of frequency parameter f with rotating speed Ω, for the different boundary conditions and initial pressure parameters p. From Figs. 4.75–4.78, it can be seen that, the variations of frequency parameters f of both backward and forward waves with circumferential wave number n are similar for the four different boundary conditions. The frequency parameters f first decrease at similar gradients for the cases associated with different initial pressure parameters p, and then increase at different gradients, depending on the magnitude of the initial pressure parameter p. By comparing these figures it is also noted that the differences in frequency parameters f between the Cs–Cl and Ss–Cl shells, or those between the Cs–Sl and Ss–Sl

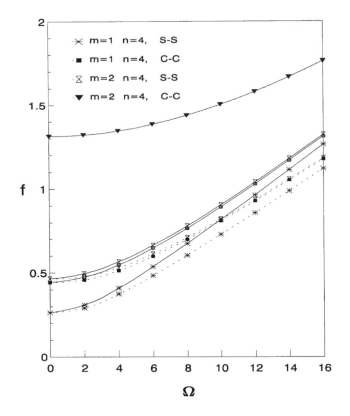

Figure 4.90 Variation of the frequency parameter *f* with
the rotating velocity Ω (rps) at different vibrational modes
for a sandwich-type conical shell ($\alpha = 45°$, $h/a = 0.008$,
$L/a = 20$).

shells, become very small when the circumferential wave number *n* is large. From
Figs. 4.79–4.82 it is also observed that, with increasing rotating speed Ω, the frequency
parameters *f* of the backward waves increase while those of the forward waves decrease so
that the frequency differences between the backward and forward waves increase
monotonically. It is also evident that the influence of boundary condition on the variation of
frequency parameter *f* with rotating speed Ω is significant at the present circumferential
wave number $n = 2$.

g)   Discussion on wave number.

The vibrational modes of a shell of revolution are defined by the meridional wave
number *m* and circumferential wave number *n*. These represent the numbers of cycles per unit

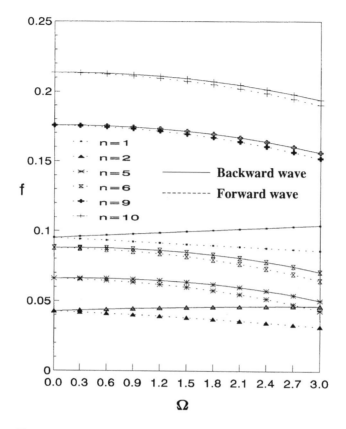

Figure 4.91   Variation of frequency parameter f with rotating speed Ω for a rotating composite conical shell with Cs–Sl boundary condition ($\alpha = 5°$, $h/a = 0.025$, $L/a = 15$, Rem = 3).

distance for the shape of a wave or waveform in the meridional and circumferential directions, respectively. For a stationary shell of revolution, the vibrational modes (m, n) consist of the standing waves whose waveform does not propagate through the shell of revolution, but whose vibrational amplitude is a function of coordinate position. When the shell of revolution starts to rotate, the standing wave motion will be transformed, and backward or forward travelling waves will emerge, depending on the rotating direction. The waveform is constant but moves through the shell of revolution. In this section, the characteristics of the meridional and circumferential wave numbers m and n are discussed in detail for the free vibration of rotating shells.

For a rotating conical shell with cone angle $\alpha = 45°$ and simply supported boundary conditions at both ends, the variations of the frequency parameter f against the rotating

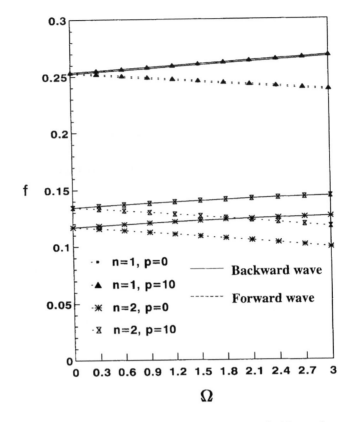

Figure 4.92 Variation of frequency parameter *f* with rotating speed Ω for a rotating conical shell with Cs–Sl boundary condition ($\alpha = 15°$, $h/a = 0.025$, $L/a = 15$).

speed Ω for the various modes (m, n) of the vibration are shown in Figs. 4.83 and 4.84. The figures show that mode $(2, 2)$ has generally higher frequencies for both the backward and forward waves than those of mode $(1, 2)$. The same trend is also observed for the respective modes $(2, 4)$ and $(1, 4)$. When the rotating speed Ω increases, the frequencies for the backward and forward waves of the four modes shown here are observed to increase quasi-linearly with the rotational speed.

When a rotating Ss–Sl sandwich-type conical shell, defined by Table 4.13, is considered, Figs. 4.85 and 4.86 are generated for the case where $h/a = 0.008$, $L/a = 20$ and cone angle $\alpha = 45°$. Figure 4.85 shows the variation of the frequency parameter *f* with the rotating velocity Ω for modes $(1, 2)$ and $(2, 2)$ while Fig. 4.86 shows that of modes $(1, 4)$ and $(2, 4)$. As in Fig. 4.83, Fig. 4.85 shows that mode $(2, 2)$ has generally higher frequency parameters *f* than mode $(1, 2)$, and correspondingly in Fig. 4.86, we note

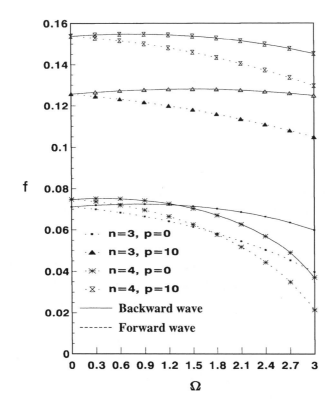

Figure 4.93 Variation of frequency parameter f with
rotating speed Ω for a rotating conical shell with Cs–Sl
boundary condition ($\alpha = 15°$, $h/a = 0.025$, $L/a = 15$).

that mode $(2, 4)$ has generally higher frequency parameters f than mode $(1, 4)$. Once
again, the frequency parameters f of both backward and forward waves of the four modes
shown here are found to increase quasi-linearly with rotating velocity Ω.

For the discussion on the coupled influence of the vibrational mode and boundary
condition on the frequency characteristics of the rotating isotropic conical shell, Figs. 4.87
and 4.88 are generated for clamped (C–C) and simply supported (S–S) boundary
conditions. The geometric properties are taken as $h/a = 0.02$, $L/a = 20$ while the cone
angle is $\alpha = 45°$. These figures show the variations of the frequency parameter f with the
rotating velocity Ω at various vibrational modes (m, n) for the C–C and S–S rotating
conical shells. From these two figures, it can be seen that the difference in the frequency
parameters between the C–C and S–S conical shells for mode $(2, 2)$ is much larger than
that of mode $(1, 2)$. Similarly, this difference in the frequency parameters for mode $(2, 4)$ is

also much larger than that of mode $(1, 4)$. For any vibrational mode (m, n), the frequency curve corresponding to the C–C shell is always larger than that of a corresponding S–S shell. In addition, it should be noted that for the S–S shell, the absolute value of the frequency parameter f of backward wave is generally larger than that of the forward wave. Also, the differences between the frequency parameters of the backward and forward waves generally increase with the increase in the rotating velocity Ω. However, for the C–C shell, similarly, trends can be observed only in the low-order vibrational mode such as $(1, 2)$ or $(1, 4)$. In the high-order vibrational mode such as $(2, 2), (2, 4)$, the difference in the frequency parameters between the backward and forward waves is minimal and tends to vanish. From the above studies, we can conclude that the boundary conditions significantly affect the variation of the frequency parameter f with the rotating velocity Ω, and such influence becomes more significant for higher order vibrational modes.

For a rotating sandwich-type conical shell, the coupled influence of vibrational mode and boundary condition is examined in Figs. 4.89 and 4.90. The material properties of each layer of this sandwich-type shell are listed in Table 4.13. These figures show the influence of vibrational mode (m, n) on the variations of frequency parameter f with rotating speed Ω, for different boundary conditions. Figure 4.89 shows that mode $(2, 2)$ has generally higher frequency parameters f than mode $(1, 2)$ for both the backward and forward waves. Moreover, the figure indicates that the influence of boundary condition for mode $(2, 2)$ is more significant than that for mode $(1, 2)$. The same characteristics are also observed between modes $(2, 4)$ and $(1, 4)$ in Fig. 4.90.

For a rotating glass fibre–epoxy laminated composite conical shell with the cone angle $\alpha = 5°$ and Cs–Sl boundary condition, Fig. 4.91 shows the influence of circumferential wave number n on the relationship between frequency parameter f and rotating speed Ω. From the figure it can be observed that the differences in frequency parameters f between the backward and forward waves decrease with increasing circumferential wave number n, while increase with increasing rotating speed Ω. At small circumferential wave numbers n, we note that the frequency parameter f of the backward wave increases and that of the forward wave decreases, with the increase in the rotating speed Ω. When the circumferential wave number n increases, the frequency parameters f of both the backward and forward waves decrease, with increasing rotating speed Ω.

To study the coupled influence of the circumferential wave number n and initial uniform pressure load on the frequency characteristics, Figs. 4.92 and 4.93 have been depicted. These figures show the variation of the frequency parameter f against the rotating speed Ω for the rotating Cs–Sl conical shell with cone angle $\alpha = 15°$, for different circumferential wave numbers n and initial pressure parameters p. It is evident from these figures that, with the increase in the circumferential wave number n, the difference in the frequency parameters f for a particular mode due to the presence of initial pressure parameter increases monotonically. Also, with the increase in the rotating speed Ω, the difference in frequency parameters f between the backward and forward waves increases monotonically. From the above discussions, we note the significant influence of circumferential wave number n on the frequency characteristics. When circumferential wave number n is small, the influence of circumferential wave number n is more significant than that of the initial uniform pressure load. When circumferential wave number n becomes larger, the influence of initial uniform pressure load becomes more significant than that of rotating speed Ω or circumferential wave number n. The influence of the rotating speed Ω on the frequency characteristics at large circumferential wave number n is more significant than that at small circumferential wave number n.

Appendix A

The differential operator matrix $\mathbf{L} = [L_{ij}]$ $(i, j = 1, 2, 3)$ in Eq. (4.33) can be written as follows:

$$L_{11} = -\frac{A_{22}\sin^2\alpha}{r^2(x)} - \rho h\frac{\partial^2}{\partial t^2} + \left(\frac{A_{66}}{r^2(x)} + \Omega^2\rho h\right)\frac{\partial^2}{\partial\theta^2}$$
$$+ \frac{A_{11}\sin\alpha}{r(x)}\frac{\partial}{\partial x} + \frac{2A_{16}}{r(x)}\frac{\partial^2}{\partial x\,\partial\theta} + A_{11}\frac{\partial^2}{\partial x^2} \tag{A.1}$$

$$L_{12} = \frac{2(B_{16}+B_{26})\cos\alpha\sin^2\alpha}{r^3(x)} + \frac{A_{26}\sin^2\alpha}{r^2(x)} + 2\Omega\rho h\sin\alpha\frac{\partial}{\partial t}$$
$$- \left(\frac{(B_{12}+B_{22}+2B_{66})\cos\alpha\sin\alpha}{r^3(x)} + \frac{(A_{22}+A_{66})\sin\alpha}{r^2(x)}\right)\frac{\partial}{\partial\theta}$$
$$+ \left(\frac{B_{26}\cos\alpha}{r^3(x)} + \frac{A_{26}}{r^2(x)}\right)\frac{\partial^2}{\partial\theta^2} - \left(\frac{2(B_{16}+B_{26})\cos\alpha\sin\alpha}{r^2(x)} + \frac{A_{26}\sin\alpha}{r(x)}\right)\frac{\partial}{\partial x}$$
$$+ \left(\frac{(B_{12}+2B_{66})\cos\alpha}{r^2(x)} + \frac{(A_{12}+A_{66})}{r(x)}\right)\frac{\partial^2}{\partial x\,\partial\theta} + \left(A_{16} + \frac{2B_{16}\cos\alpha}{r(x)}\right)\frac{\partial^2}{\partial x^2} \tag{A.2}$$

$$L_{13} = -\frac{A_{22}\cos\alpha\sin\alpha}{r^2(x)} - \left(\frac{2(B_{16}+B_{26})\sin^2\alpha}{r^3(x)} - \frac{A_{26}\cos\alpha}{r^2(x)}\right)\frac{\partial}{\partial\theta}$$

$$+ \frac{(B_{12}+B_{22}+2B_{66})\sin\alpha}{r^3(x)}\frac{\partial^2}{\partial\theta^2} - \frac{B_{26}}{r^3(x)}\frac{\partial^3}{\partial\theta^3}$$

$$+ \left(\frac{B_{22}\sin^2\alpha}{r^2(x)} + \frac{A_{12}\cos\alpha}{r(x)} - \Omega^2\rho h r(x)\cos\alpha\right)\frac{\partial}{\partial x} + \frac{(2B_{16}+B_{26})\sin\alpha}{r^2(x)}\frac{\partial^2}{\partial x\,\partial\theta}$$

$$- \frac{(B_{12}+2B_{66})}{r^2(x)}\frac{\partial^3}{\partial x\,\partial\theta^2} - \frac{B_{11}\sin\alpha}{r(x)}\frac{\partial^2}{\partial x^2} - \frac{3B_{16}}{r(x)}\frac{\partial^3}{\partial x^2\,\partial\theta} - B_{11}\frac{\partial^3}{\partial x^3} \tag{A.3}$$

$$L_{21} = -\frac{B_{26}\cos\alpha\sin^2\alpha}{r^3(x)} + \frac{A_{26}\sin^2\alpha}{r^2(x)} - 2\Omega\rho h\sin\alpha\frac{\partial}{\partial t}$$

$$+\left(\frac{(B_{22}-B_{66})\cos\alpha\sin\alpha}{r^3(x)} + \frac{(A_{22}+A_{66})\sin\alpha}{r^2(x)} + \Omega^2\rho h\sin\alpha\right)\frac{\partial}{\partial\theta}$$

$$+\left(\frac{B_{26}\cos\alpha}{r^3(x)} + \frac{A_{26}}{r^2(x)}\right)\frac{\partial^2}{\partial\theta^2} + \left(\frac{B_{26}\cos\alpha\sin\alpha}{r^2(x)} + \frac{(2A_{16}+A_{26})\sin\alpha}{r(x)}\right)\frac{\partial}{\partial x}$$

$$+\left(\Omega^2\rho hr(x) + \frac{(B_{12}+B_{66})\cos\alpha}{r^2(x)} + \frac{(A_{12}+A_{66})}{r(x)}\right)\frac{\partial^2}{\partial x\,\partial\theta} + \left(A_{16} + \frac{B_{16}\cos\alpha}{r(x)}\right)\frac{\partial^2}{\partial x^2}$$

$$(A.4)$$

$$L_{22} = \frac{4D_{66}\cos^2\alpha\sin^2\alpha}{r^4(x)} + \frac{B_{66}\cos\alpha\sin^2\alpha}{r^3(x)} - \frac{A_{66}\sin^2\alpha}{r^2(x)} - \rho h\frac{\partial^2}{\partial t^2}$$

$$-\left(\frac{4D_{26}\cos^2\alpha\sin\alpha}{r^4(x)} + \frac{4B_{26}\cos\alpha\sin\alpha}{r^3(x)}\right)\frac{\partial}{\partial\theta}$$

$$+\left(\frac{D_{22}\cos^2\alpha}{r^4(x)} + \frac{2B_{22}\cos\alpha}{r^3(x)} + \frac{A_{22}}{r^2(x)}\right)\frac{\partial^2}{\partial\theta^2}$$

$$-\left(\frac{4D_{66}\cos^2\alpha\sin\alpha}{r^3(x)} + \frac{B_{66}\cos\alpha\sin\alpha}{r^2(x)} - \frac{A_{66}\sin\alpha}{r(x)} - \Omega^2\rho hr(x)\sin\alpha\right)\frac{\partial}{\partial x}$$

$$+\left(\frac{3D_{26}\cos^2\alpha}{r^3(x)} + \frac{5B_{26}\cos\alpha}{r^2(x)} + \frac{2A_{26}}{r(x)}\right)\frac{\partial^2}{\partial x\,\partial\theta}$$

$$+\left(\frac{2D_{66}\cos^2\alpha}{r^2(x)} + \frac{3B_{66}\cos\alpha}{r(x)} + A_{66}\right)\frac{\partial^2}{\partial x^2}$$

$$(A.5)$$

$$L_{23} = -\frac{B_{26}\cos^2\alpha\sin\alpha}{r^3(x)} + \frac{A_{26}\cos\alpha\sin\alpha}{r^2(x)} - 2\Omega\rho h\cos\alpha\frac{\partial}{\partial t}$$

$$-\left(\frac{4D_{66}\cos\alpha\sin^2\alpha}{r^4(x)} - \frac{B_{22}\cos^2\alpha}{r^3(x)} - \frac{A_{22}\cos\alpha}{r^2(x)}\right)\frac{\partial}{\partial\theta}$$

$$+\left(\frac{4D_{26}\cos\alpha\sin\alpha}{r^4(x)} + \frac{2B_{26}\sin\alpha}{r^3(x)}\right)\frac{\partial^2}{\partial\theta^2} - \left(\frac{D_{22}\cos\alpha}{r^4(x)} + \frac{B_{22}}{r^3(x)}\right)\frac{\partial^3}{\partial\theta^3}$$

$$+\left(\frac{D_{26}\cos\alpha\sin^2\alpha}{r^3(x)} + \frac{B_{26}(\cos^2\alpha-\sin^2\alpha)}{r^2(x)} + \frac{A_{26}\cos\alpha}{r(x)}\right)\frac{\partial}{\partial x}$$

$$-\left(\frac{(D_{22}-4D_{66})\cos\alpha\sin\alpha}{r^3(x)} + \frac{B_{22}\sin\alpha}{r^2(x)}\right)\frac{\partial^2}{\partial x\,\partial\theta}$$

$$-\left(\frac{3D_{26}\cos\alpha}{r^3(x)}+\frac{3B_{26}}{r^2(x)}\right)\frac{\partial^3}{\partial x\,\partial\theta^2}-\left(\frac{D_{26}\cos\alpha\sin\alpha}{r^2(x)}+\frac{(2B_{16}+B_{26})\sin\alpha}{r(x)}\right)\frac{\partial^2}{\partial x^2}$$

$$-\left(\frac{(D_{12}+2D_{66})\cos\alpha}{r^2(x)}+\frac{(B_{12}+2B_{66})}{r(x)}\right)\frac{\partial^3}{\partial x^2\,\partial\theta}-\left(B_{16}+\frac{D_{16}\cos\alpha}{r(x)}\right)\frac{\partial^3}{\partial x^3}\qquad\text{(A.6)}$$

$$L_{31}=\frac{B_{22}\sin^3\alpha}{r^3(x)}-\frac{A_{22}\cos\alpha\sin\alpha}{r^2(x)}+\Omega^2\rho h\cos\alpha\sin\alpha$$

$$-\left(\frac{B_{26}\sin^2\alpha}{r^3(x)}+\frac{A_{26}\cos\alpha}{r^2(x)}\right)\frac{\partial}{\partial\theta}+\frac{(B_{22}-2B_{66})\sin\alpha}{r^3(x)}\frac{\partial^2}{\partial\theta^2}+\frac{B_{26}}{r^3(x)}\frac{\partial^3}{\partial\theta^3}$$

$$-\left(\frac{B_{22}\sin^2\alpha}{r^2(x)}+\frac{A_{12}\cos\alpha}{r(x)}+\Omega^2\rho hr(x)\cos\alpha\right)\frac{\partial}{\partial x}+\frac{B_{26}\sin\alpha}{r^2(x)}\frac{\partial^2}{\partial x\,\partial\theta}$$

$$+\frac{(B_{12}+2B_{66})}{r^2(x)}\frac{\partial^3}{\partial x\,\partial\theta^2}+\frac{2B_{11}\sin\alpha}{r(x)}\frac{\partial^2}{\partial x^2}+\frac{3B_{16}}{r(x)}\frac{\partial^3}{\partial x^2\,\partial\theta}+B_{11}\frac{\partial^3}{\partial x^3}\qquad\text{(A.7)}$$

$$L_{32}=-\frac{4(D_{16}+D_{26})\cos\alpha\sin^3\alpha}{r^4(x)}+\frac{(2\cos^2\alpha-\sin^2\alpha)B_{26}\sin\alpha}{r^3(x)}+\frac{A_{26}\cos\alpha\sin\alpha}{r^2(x)}$$

$$+2\Omega\rho h\cos\alpha\frac{\partial}{\partial t}-\left(\frac{6D_{26}\cos\alpha\sin\alpha}{r^4(x)}+\frac{3B_{26}\sin\alpha}{r^3(x)}\right)\frac{\partial^2}{\partial\theta^2}$$

$$+\left(\frac{D_{22}\cos\alpha}{r^4(x)}+\frac{B_{22}}{r^3(x)}\right)\frac{\partial^3}{\partial\theta^3}+\left(\frac{2(D_{12}+D_{22}+4D_{66})\cos\alpha\sin^2\alpha}{r^4(x)}\right.$$

$$+\left.\frac{(\sin^2\alpha-\cos^2\alpha)B_{22}+2B_{66}\sin^2\alpha}{r^3(x)}-\frac{A_{22}\cos\alpha}{r^2(x)}\right)\frac{\partial}{\partial\theta}$$

$$+\left(\frac{4(D_{16}+D_{26})\cos\alpha\sin^2\alpha}{r^3(x)}+\frac{(\sin^2\alpha-2\cos^2\alpha)B_{26}}{r^2(x)}-\frac{A_{26}\cos\alpha}{r(x)}\right)\frac{\partial}{\partial x}$$

$$-\left(\frac{(2D_{12}+D_{22}+8D_{66})\cos\alpha\sin\alpha}{r^3(x)}+\frac{(B_{22}+2B_{66})\sin\alpha}{r^2(x)}\right)\frac{\partial^2}{\partial x\,\partial\theta}$$

$$+\left(\frac{4D_{26}\cos\alpha}{r^3(x)}+\frac{3B_{26}}{r^2(x)}\right)\frac{\partial^3}{\partial x\,\partial\theta^2}-\left(\frac{2(D_{16}+D_{26})\cos\alpha\sin\alpha}{r^2(x)}\right.$$

$$+\left.\frac{(B_{26}-B_{16})\sin\alpha}{r(x)}\right)\frac{\partial^2}{\partial x^2}+\left(\frac{(D_{12}+4D_{66})\cos\alpha}{r^2(x)}+\frac{(B_{12}+2B_{66})}{r(x)}\right)\frac{\partial^3}{\partial x^2\,\partial\theta}$$

$$+\left(B_{16}+\frac{2D_{16}\cos\alpha}{r(x)}\right)\frac{\partial^3}{\partial x^3}\qquad\text{(A.8)}$$

$$L_{33} = \frac{B_{22}\cos\alpha\sin^2\alpha}{r^3(x)} - \frac{A_{22}\cos^2\alpha}{r^2(x)} + \Omega^2\rho h\cos^2\alpha - \rho h\frac{\partial^2}{\partial t^2}$$

$$+ \left(\frac{4(D_{16}+D_{26})\sin^3\alpha}{r^4(x)} - \frac{4B_{26}\cos\alpha\sin\alpha}{r^3(x)}\right)\frac{\partial}{\partial\theta} + \frac{6D_{26}\sin\alpha}{r^4(x)}\frac{\partial^3}{\partial\theta^3} - \frac{D_{22}}{r^4(x)}$$

$$\times\frac{\partial^4}{\partial\theta^4} - \left(\frac{2(D_{12}+D_{22}+4D_{66})\sin^2\alpha}{r^4(x)} + \frac{2B_{22}\cos\alpha}{r^3(x)} + \Omega^2\rho h\right)\frac{\partial^2}{\partial\theta^2}$$

$$- \frac{D_{22}\sin^3\alpha}{r^3(x)}\frac{\partial}{\partial x} - \left(\frac{2(2D_{16}+D_{26})\sin^2\alpha}{r^3(x)} - \frac{4B_{26}\cos\alpha}{r^2(x)}\right)\frac{\partial^2}{\partial x\,\partial\theta}$$

$$+ \frac{2(D_{12}+4D_{66})\sin\alpha}{r^3(x)}\frac{\partial^3}{\partial x\,\partial\theta^2} - \frac{4D_{26}}{r^3(x)}\frac{\partial^4}{\partial x\,\partial\theta^3}$$

$$+ \left(\frac{D_{22}\sin^2\alpha}{r^2(x)} + \frac{2B_{12}\cos\alpha}{r(x)}\right)\frac{\partial^2}{\partial x^2} + \frac{2D_{16}\sin\alpha}{r^2(x)}\frac{\partial^3}{\partial x^2\,\partial\theta} - \frac{(2D_{12}+4D_{66})}{r^2(x)}$$

$$\times\frac{\partial^4}{\partial x^2\,\partial\theta^2} - \frac{2D_{11}\sin\alpha}{r(x)}\frac{\partial^3}{\partial x^3} - \frac{4D_{16}}{r(x)}\frac{\partial^4}{\partial x^3\,\partial\theta} - D_{11}\frac{\partial^4}{\partial x^4} \tag{A.9}$$

Appendix B

The differential operator matrix $\mathbf{L}^* = [L_{ij}^*]$ $(i, j = 1, 2, 3)$ in Eq. (4.36) can be written as follows:

$$L_{11}^* = A_{11}\frac{d^2}{dx^2} + \frac{A_{11}\sin\alpha}{r(x)}\frac{d}{dx} - \frac{A_{22}\sin^2\alpha + n^2A_{66}}{r^2(x)} - \rho h(n^2\Omega^2 - \omega^2) \tag{B.1}$$

$$L_{12}^* = \left(\frac{n(A_{12}+A_{66})}{r(x)} + \frac{n(B_{12}+2B_{66})\cos\alpha}{r^2(x)}\right)\frac{d}{dx} + 2\Omega\omega\rho h\sin\alpha$$

$$- \frac{n(A_{22}+A_{66})\sin\alpha}{r^2(x)} - \frac{n(B_{12}+B_{22}+2B_{66})\cos\alpha\sin\alpha}{r^3(x)} \tag{B.2}$$

$$L_{13}^* = -B_{11}\frac{d^3}{dx^3} - \frac{B_{11}\sin\alpha}{r(x)}\frac{d^2}{dx^2} - \frac{A_{22}\cos\alpha\sin\alpha}{r^2(x)} - \frac{n^2(B_{12}+B_{22}+2B_{66})\sin\alpha}{r^3(x)}$$

$$+ \left(\frac{A_{12}\cos\alpha}{r(x)} + \frac{n^2(B_{12}+2B_{66})+B_{22}\sin^2\alpha}{r^2(x)} - \Omega^2\rho hr(x)\cos\alpha\right)\frac{d}{dx} \tag{B.3}$$

$$L_{21}^* = -n\left(\frac{A_{12} + A_{66}}{r(x)} + \frac{(B_{12} + B_{66})\cos\alpha}{r^2(x)} + \Omega^2\rho h r(x)\right)\frac{d}{dx} - \Omega\rho h \sin\alpha(n\Omega$$

$$- 2\omega) - \frac{n(A_{22} + A_{66})\sin\alpha}{r^2(x)} - \frac{n(B_{22} - B_{66})\cos\alpha\sin\alpha}{r^3(x)} \tag{B.4}$$

$$L_{22}^* = \left(A_{66} + \frac{3B_{66}\cos\alpha}{r(x)} + \frac{2D_{66}\cos^2\alpha}{r^2(x)}\right)\frac{d^2}{dx^2} + \rho h\omega^2$$

$$+ \left(\frac{A_{66}\sin\alpha}{r(x)} - \frac{B_{66}\cos\alpha\sin\alpha}{r^2(x)} - \frac{4D_{66}\cos^2\alpha\sin\alpha}{r^3(x)} + \Omega^2\rho h r(x)\sin\alpha\right)\frac{d}{dx}$$

$$- \frac{n^2A_{22} + A_{66}\sin^2\alpha}{r^2(x)} - \frac{(2n^2B_{22} - B_{66}\sin^2\alpha)\cos\alpha}{r^3(x)} - \frac{(n^2D_{22} - 4D_{66}\sin^2\alpha)\cos^2\alpha}{r^4(x)} \tag{B.5}$$

$$L_{23}^* = n\left(\frac{B_{12} + 2B_{66}}{r(x)} + \frac{(D_{12} + 2D_{66})\cos\alpha}{r^2(x)}\right)\frac{d^2}{dx^2} + 2\Omega\omega\rho h \cos\alpha$$

$$+ n\left(\frac{B_{22}\sin\alpha}{r^2(x)} + \frac{(D_{22} - 4D_{66})\cos\alpha\sin\alpha}{r^3(x)}\right)\frac{d}{dx} - \frac{nA_{22}\cos\alpha}{r^2(x)}$$

$$- \frac{nB_{22}(\cos^2\alpha + n^2)}{r^3(x)} + \frac{4nD_{66}\cos\alpha\sin^2\alpha - n^3D_{22}\cos\alpha}{r^4(x)} \tag{B.6}$$

$$L_{31}^* = B_{11}\frac{d^3}{dx^3} + \frac{2B_{11}\sin\alpha}{r(x)}\frac{d^2}{dx^2} + \Omega^2\rho h\cos\alpha\sin\alpha$$

$$- \left(\frac{A_{12}\cos\alpha}{r(x)} + \frac{n^2(B_{12} + 2B_{66}) + B_{22}\sin^2\alpha}{r^2(x)} + \Omega^2\rho h r(x)\cos\alpha\right)\frac{d}{dx}$$

$$- \frac{A_{22}\cos\alpha\sin\alpha}{r^2(x)} - \frac{B_{22}\sin\alpha(n^2 - \sin^2\alpha) - 2n^2B_{66}\sin\alpha}{r^3(x)} \tag{B.7}$$

$$L_{32}^* = n\left(\frac{B_{12} + 2B_{66}}{r(x)} + \frac{(D_{12} + 4D_{66})\cos\alpha}{r^2(x)}\right)\frac{d^2}{dx^2} + 2\Omega\omega\rho h \cos\alpha$$

$$- n\left(\frac{(B_{22} + 2B_{66})\sin\alpha}{r^2(x)} + \frac{(2D_{12} + D_{22} + 8D_{66})\cos\alpha\sin\alpha}{r^3(x)}\right)\frac{d}{dx}$$

$$- \frac{nA_{22}\cos\alpha}{r^2(x)} + \frac{nB_{22}(\sin^2\alpha - \cos^2\alpha - n^2) + 2nB_{66}\sin^2\alpha}{r^3(x)}$$

$$+ \frac{2n(D_{12} + D_{22} + 4D_{66})\cos\alpha\sin^2\alpha - n^3D_{22}\cos\alpha}{r^4(x)} \tag{B.8}$$

$$L_{33}^* = -D_{11}\frac{d^4}{dx^4} - \frac{2D_{11}\sin\alpha}{r(x)}\frac{d^3}{dx^3}$$

$$+ \left(\frac{2B_{12}\cos\alpha}{r(x)} + \frac{n^2(2D_{12}+4D_{66})+D_{22}\sin^2\alpha}{r^2(x)}\right)\frac{d^2}{dx^2}$$

$$- \frac{2n^2(D_{12}+4D_{66})\sin\alpha + D_{22}\sin^3\alpha}{r^3(x)}\frac{d}{dx} + \rho h(\Omega^2\cos^2\alpha + \omega^2 + \Omega^2 n^2)$$

$$- \frac{A_{22}\cos^2\alpha}{r^2(x)} + \frac{B_{22}\cos\alpha(\sin^2\alpha + 2n^2)}{r^3(x)}$$

$$+ \frac{2n^2(D_{12}+D_{22}+4D_{66})\sin^2\alpha - n^4D_{22}}{r^4(x)} \qquad\qquad (B.9)$$

Chapter 5
Free Vibration of Thick Rotating Cylindrical Shells

5.1   Introduction.

Although the dynamic behavior of rotating shells has been studied for over a century, a thorough search of the literature will reveal that most of the published works are based on thin-shell theories. Researchers have found that the application of a classical thin-shell theory to the thick shells can lead to significant errors in natural frequency analysis. Classical thin-shell theories are unable to achieve sufficient accuracy when applied to thick shells. Thick shells have a number of distinctly different features from thin shells. One of these features is that the transverse shear deformation can no longer be neglected in thick shells. In many loading cases, the radial stress distribution of thick shells becomes very important and needs to be incorporated into the formulations. Additionally, in the analysis of thick shells, initial curvatures do not only contribute to the stress and moment resultants, but also result in nonlinear distributions of in-plane stresses along the thickness direction. Thus, proper thick-shell theories are required to refine the Kirchhoff hypotheses.

Some early thick-shell theories incorporated the classic nonlinear kinematic strain–displacement relations. Mindlin *et al.* [1951, 1954, 1956] developed the first-order shear deformation theory (FSDT) requiring a shear correction factor to compensate for the error due to the assumption of a constant shear strain resulting in constant shear stress through the thickness, and thus violating the zero shear stress conditions at the free surfaces. The value of the shear correction factor used often depends on the material properties and geometrical parameters as well as loading and boundary conditions. Reddy [1984a,b, 1986, 1997, 1999, 2004] proposed the third-order shear deformation theory (TSDT) to refine the zero shear stress condition at the free surface. Even higher order theories have been developed, but the formulations become much more complicated and the relatively small improvements in computational accuracy render them unfeasible for practical use.

Although various thick-shell theories have been developed, studies on the dynamics of thick rotating shells are very limited. One of these is the work carried out by Sivadas & Ganesan [1994] who examined the influences of rotation and damping factor on the free vibration of a thick rotating circular cylindrical shell. In that work, the finite element method and moderately thick shell theory with consideration of shear deformation and rotary inertia were used. Sivadas [1995] also employed similar method and theory for the free vibration analysis of thick rotating circular conical shell with pre-stress and damping effects. For the free vibration of thick rotating laminated composite cylindrical shells, Lam & Wu [1999] investigated the effects of rotating velocity, circumferential wave number and the length and thickness ratios. Guo *et al.* [2001] contributed a very interesting work on the vibration analysis of thick rotating cylindrical shells based on a three-dimensional elasticity model.

In this chapter, based on the first-order Mindlin shell theory, a frequency analysis is carried out for the free vibration of thick rotating circular cylindrical shells. The differences in frequency characteristics between thin and thick rotating cylindrical shells are examined, and the influences of various physical and geometrical parameters are discussed in detail. In addition, a three-dimensional vibration analysis of thick rotating circular cylindrical shells is also presented using the finite element formulation with consideration of classic nonlinear kinematics.

5.2   Natural frequency analysis by Mindlin shell theory.

Since Mindlin proposed the FSDT in 1951, it has been applied to various thick plate and shell analyses for the accounting of transverse shear effects. Compared with classic nonlinear kinematics and higher order shear deformation theories, the FSDT combines the advantages of acceptable accuracy and lower computational efforts. In this section, Mindlin's FSDT is applied to the free vibration of thick rotating simply supported laminated composite cylindrical shells. The differences in frequency characteristics between thin and thick rotating cylindrical shells are examined. Discussions are also made for the influences of rotating velocity, circumferential wave number and geometric thickness and length ratios (H/R and L/R) on the frequency characteristics of the thick rotating cylindrical shells.

a)   *Rotating Mindlin shell theory — development.*

Consider a thick circular laminated composite cylindrical shell rotating about its symmetrical horizontal axis at a constant angular velocity Ω, such as that shown earlier in Fig. 3.1. In a similar manner, the reference surface for the deformations of the rotating cylindrical shell is taken to be the middle surface on which an orthogonal coordinate system (x, θ, z) is fixed. L is the length and R the constant mean radius. The thickness of the cylindrical shell is denoted here by H. Deformations of the thick rotating cylindrical shell are defined by u, v and w in the longitudinal x, circumferential θ and radial z directions, respectively. Rotations of the transverse normal about the θ and x axes are denoted by ψ_x and ψ_θ and can be written as

$$\psi_x = \frac{\partial u}{\partial z}, \qquad \psi_\theta = \frac{\partial v}{\partial z}. \tag{5.1}$$

In the FSDT, the Kirchhoff hypothesis is relaxed such that the transverse normals are no longer constrained to remain perpendicular to the middle surface after deformation. This amounts to the inclusion of transverse shear strains in the theoretical formulations. The inextensibility of the transverse normals requires that the displacement w be independent of the thickness coordinate z. The displacement field of the FSDT is of the form

$$u(x, \theta, z, t) = u_0(x, \theta, t) + z\psi_x(x, \theta, t) \tag{5.2}$$

$$v(x, \theta, z, t) = v_0(x, \theta, t) + z\psi_\theta(x, \theta, t) \tag{5.3}$$

$$w(x, \theta, z, t) = w_0(x, \theta, t) \tag{5.4}$$

where u_0, v_0, w_0, ψ_x and ψ_θ are unknown functions to be solved, in which u_0, v_0 and w_0 are the displacements of a point on the middle surface $z = 0$.

The corresponding kinematic strain–displacement relations associated with the displacement field equations (5.2)–(5.4), and shown earlier in Eqs. (2.18)–(2.22), are rewritten here as

$$\varepsilon_x = e_x + z\kappa_x = \frac{\partial u}{\partial x} + z\frac{\partial \psi_x}{\partial x} \tag{5.5}$$

$$\varepsilon_\theta = e_\theta + z\kappa_\theta = \frac{1}{R}\frac{\partial v}{\partial \theta} + \frac{w}{R} + \frac{z}{R}\frac{\partial \psi_\theta}{\partial \theta} \tag{5.6}$$

$$\varepsilon_{x\theta} = e_{x\theta} + z\kappa_{x\theta} = \frac{\partial v}{\partial x} + \frac{1}{R}\frac{\partial u}{\partial \theta} + z\left(\frac{\partial \psi_\theta}{\partial x} + \frac{1}{R}\frac{\partial \psi_x}{\partial \theta} + \frac{1}{2R}\left(\frac{\partial v}{\partial x} - \frac{1}{R}\frac{\partial u}{\partial \theta}\right)\right) \tag{5.7}$$

$$\varepsilon_{xz} = \frac{\partial w}{\partial x} - \frac{v}{R} + \psi_x \tag{5.8}$$

$$\varepsilon_{\theta z} = \frac{1}{R}\frac{\partial w}{\partial \theta} + \psi_\theta. \tag{5.9}$$

Using Hamilton's principle and accounting for the Coriolis acceleration, the governing equations of motion in terms of the in-plane force and moment resultants are derived as

$$\frac{\partial N_x}{\partial x} + \frac{\partial N_{\theta x}}{R\partial\theta} = I_0 \frac{\partial^2 u_0}{\partial t^2} + I_1 \frac{\partial^2 \psi_x}{\partial t^2} \tag{5.10}$$

$$\frac{\partial N_{x\theta}}{\partial x} + \frac{\partial N_\theta}{R\partial\theta} = I_0 \frac{\partial^2 v_0}{\partial t^2} + I_1 \frac{\partial^2 \psi_\theta}{\partial t^2} + 2I_0\Omega \frac{\partial w_0}{\partial t} - I_0\Omega^2 v_0 \tag{5.11}$$

$$\frac{\partial Q_x}{\partial x} + \frac{\partial Q_\theta}{R\partial\theta} - \frac{N_\theta}{R} = I_0 \frac{\partial^2 w_0}{\partial t^2} - 2I_0\Omega \frac{\partial v_0}{\partial t} - I_0\Omega^2 w_0 \tag{5.12}$$

$$\frac{\partial M_x}{\partial x} + \frac{\partial M_{\theta x}}{R\partial\theta} - Q_x = I_1 \frac{\partial^2 u_0}{\partial t^2} + I_2 \frac{\partial^2 \psi_x}{\partial t^2} \tag{5.13}$$

$$\frac{\partial M_{x\theta}}{\partial x} + \frac{\partial M_\theta}{R\partial\theta} - Q_\theta = I_1 \frac{\partial^2 v_0}{\partial t^2} + I_2 \frac{\partial^2 \psi_\theta}{\partial t^2} \tag{5.14}$$

where I_0, I_1 and I_2 are the mass moments of inertia and defined as

$$\{I_0, I_1, I_2\} = \int_{-H/2}^{H/2} \{1, z, z^2\}\rho dz. \tag{5.15}$$

For the presently studied thick laminated composite cylindrical shell, we shall assume that each layer is orthotropic and is stacked in a cross-ply manner. As a result, the coupling stiffness matrix $\mathbf{B} = [B_{ij}] = 0$, $A_{16} = A_{26} = 0$ in the extension stiffness matrix $\mathbf{A} = [A_{ij}]$, and $D_{16} = D_{26} = 0$ in the bending stiffness matrix $\mathbf{D} = [D_{ij}]$, which arise from the constitutive relation (2.34) of the cylindrical shell. Thus, the in-plane force and moment resultants can be expressed in terms of the displacements and rotations as

$$N_x = A_{11}\frac{\partial u_0}{\partial x} + A_{12}\left(\frac{w_0}{R} + \frac{\partial v_0}{R\partial\theta}\right) + \frac{D_{11}}{R}\frac{\partial\psi_x}{\partial x} + \frac{D_{12}}{R}\frac{\partial\psi_\theta}{R\partial\theta} \tag{5.16}$$

$$N_\theta = A_{12}\frac{\partial u_0}{\partial x} + A_{22}\left(\frac{w_0}{R} + \frac{\partial v_0}{R\partial\theta}\right) \tag{5.17}$$

$$N_{x\theta} = A_{66}\left(\frac{\partial v_0}{\partial x} + \frac{\partial u_0}{R\partial\theta}\right) + \frac{D_{66}}{R}\left(\frac{\partial\psi_\theta}{\partial x} + \frac{\partial\psi_x}{R\partial\theta}\right) \tag{5.18}$$

$$N_{\theta x} = A_{66}\left(\frac{\partial v_0}{\partial x} + \frac{\partial u_0}{R\partial\theta}\right) \tag{5.19}$$

$$M_x = D_{11}\left(\frac{\partial \psi_x}{\partial x} + \frac{1}{R}\frac{\partial u_0}{\partial x}\right) + \frac{D_{12}}{R}\left(\frac{w_0}{R} + \frac{\partial v_0}{R\partial \theta} + \frac{\partial \psi_\theta}{\partial \theta}\right) \tag{5.20}$$

$$M_\theta = D_{12}\frac{\partial \psi_x}{\partial x} + \frac{D_{22}}{R}\frac{\partial \psi_\theta}{\partial \theta} \tag{5.21}$$

$$M_{x\theta} = D_{66}\left(\frac{\partial \psi_\theta}{\partial x} + \frac{\partial \psi_x}{R\partial \theta}\right) + \frac{D_{66}}{R}\left(\frac{\partial v_0}{\partial x} + \frac{\partial u_0}{R\partial \theta}\right) \tag{5.22}$$

$$M_{\theta x} = D_{66}\left(\frac{\partial \psi_\theta}{\partial x} + \frac{\partial \psi_x}{R\partial \theta}\right) \tag{5.23}$$

$$Q_x = KA_{55}\left(\psi_x + \frac{\partial w_0}{\partial x}\right) \tag{5.24}$$

$$Q_\theta = KA_{44}\left(\psi_\theta + \frac{\partial w_0}{R\partial \theta}\right). \tag{5.25}$$

Substitution of the above resulting force and moment resultants, Eqs. (5.16)–(5.25), into the governing equations (5.10)–(5.14), the governing equations of motion are finally derived in the following matrix form

$$\tilde{\mathbf{L}}_{5\times5}\tilde{\mathbf{U}}_{5\times1} = 0 \tag{5.26}$$

where $\tilde{\mathbf{U}}_{5\times1} = \{u_0, v_0, w_0, \psi_x, \psi_\theta\}^{\mathrm{T}}$ is the unknown displacement-field vector to be determined. $\tilde{\mathbf{L}}_{5\times5} = [L_{ij}]$ $(i,j = 1, 2, ...5)$ is a 5×5 differential operator matrix of the unknown $\tilde{\mathbf{U}}$.

It is noted that the resulting governing equations of motion, Eq. (5.26), is general and without reference to any particular boundary condition. When the S3-type of simply supported boundary condition is considered along both edges (see Nosier & Reddy, 1992), namely

$$M_x = N_x = v_0 = w_0 = \psi_\theta = 0 \qquad \text{at} \qquad x = 0 \text{ and } L \tag{5.27}$$

the trial functions describing the displacement field for the free vibration of the thick rotating cylindrical shell may be expressed as follows

$$u_0 = \tilde{A}\cos\left(\frac{m\pi x}{L}\right)\cos(n\theta + \omega t) \tag{5.28}$$

$$v_0 = \tilde{B} \sin\left(\frac{m\pi x}{L}\right)\sin(n\theta + \omega t) \tag{5.29}$$

$$w_0 = \tilde{C} \sin\left(\frac{m\pi x}{L}\right)\cos(n\theta + \omega t) \tag{5.30}$$

$$\psi_x = \tilde{D} \cos\left(\frac{m\pi x}{L}\right)\cos(n\theta + \omega t) \tag{5.31}$$

$$\psi_\theta = \tilde{E} \sin\left(\frac{m\pi x}{L}\right)\sin(n\theta + \omega t) \tag{5.32}$$

in which $\tilde{A}$, $\tilde{B}$, $\tilde{C}$, $\tilde{D}$ and $\tilde{E}$ are the unknown vibrational amplitudes, m and n are, respectively, longitudinal and circumferential wave numbers, and ω is the natural circular frequency (rad/s) of the shell.

Substituting the trial functions of the displacement field, Eqs. (5.28)–(5.32), into the governing equations of motion, Eq. (5.26), we arrive at a characteristic matrix

$$\begin{bmatrix} \hat{C}_{11} & \hat{C}_{12} & \hat{C}_{13} & \hat{C}_{14} & \hat{C}_{15} \\ \hat{C}_{21} & \hat{C}_{22} & \hat{C}_{23} & \hat{C}_{24} & \hat{C}_{25} \\ \hat{C}_{31} & \hat{C}_{32} & \hat{C}_{33} & \hat{C}_{34} & \hat{C}_{35} \\ \hat{C}_{41} & \hat{C}_{42} & \hat{C}_{43} & \hat{C}_{44} & \hat{C}_{45} \\ \hat{C}_{51} & \hat{C}_{52} & \hat{C}_{53} & \hat{C}_{54} & \hat{C}_{55} \end{bmatrix} \begin{Bmatrix} \tilde{A} \\ \tilde{B} \\ \tilde{C} \\ \tilde{D} \\ \tilde{E} \end{Bmatrix} = \begin{Bmatrix} 0 \\ 0 \\ 0 \\ 0 \\ 0 \end{Bmatrix}. \tag{5.33}$$

Imposing the nontrivial solution condition, and then setting the determinant of the above characteristic matrix equal to zero, the eigenvalue equation with respect to the natural circular frequency ω is obtained by

$$\begin{vmatrix} \hat{C}_{11} & \hat{C}_{12} & \hat{C}_{13} & \hat{C}_{14} & \hat{C}_{15} \\ \hat{C}_{21} & \hat{C}_{22} & \hat{C}_{23} & \hat{C}_{24} & \hat{C}_{25} \\ \hat{C}_{31} & \hat{C}_{32} & \hat{C}_{33} & \hat{C}_{34} & \hat{C}_{35} \\ \hat{C}_{41} & \hat{C}_{42} & \hat{C}_{43} & \hat{C}_{44} & \hat{C}_{45} \\ \hat{C}_{51} & \hat{C}_{52} & \hat{C}_{53} & \hat{C}_{54} & \hat{C}_{55} \end{vmatrix} = 0. \tag{5.34}$$

By expanding the above determinant, Eq. (5.34), a tenth-order polynomial equation with respect to the natural circular frequency ω is obtained. If the shell is stationary ($\Omega = 0$), it will be reduced to a fifth-order polynomial equation in ω^2. When the cylindrical shell rotates at constant angular velocity, the eigensolutions of the problem include the natural circular frequencies corresponding to the backward and forward traveling waves of the thick rotating cylindrical shell.

b)   *Numerical validation and comparison.*

To validate the computational accuracy of the present theoretical formulations, four comparison studies, which compare present results with those available in the open literature, are carried out. The first two comparisons are shown in Tables 5.1 and 5.2 for two stationary cylindrical shells, namely a long thin isotropic cylindrical shell and a short thick simply supported laminated composite cylindrical shell, respectively. The other two comparisons are shown in Tables 5.3 and 5.4 for two rotating cylindrical shells, namely a long thick cylindrical shell and a short thin simply supported laminated composite cylindrical shell, respectively. It is noted that, in all the four comparisons, the shear correction factor is taken as $K = 5/6$ in the present implementation of the Mindlin shell theory.

Table 5.1

Comparison of frequency parameter $f = \omega R \sqrt{(1 - \mu^2)\rho/E}$ for a long nonrotating thin isotropic cylindrical shell ($m = 1$, $\mu = 0.3$, $H/R = 0.002$, $L/R = 200$).

n	Chen et al. [1993][a]	Soedel [1993][b]	Present
2	0.00154919	0.00200014	0.00206636
3	0.00438178	0.00489912	0.00492965
4	0.00840168	0.00894441	0.00896183
5	0.0135873	0.0141423	0.0141533
6	0.0199323	0.0204944	0.0205014
7	0.0274343	0.0280001	0.028005
8	0.0360922	0.0366607	0.0366635

[a] Calculated from equation from Chen et al. [1993]:

$$\omega_{b,f} = \frac{2n}{n^2 + 1}\Omega \pm \sqrt{\frac{n^2(n^2 - 1)^2}{n^2 + 1}\frac{EH^2}{12(1 - \mu^2)\rho R^2} + \frac{n^4 + 3}{(n^2 + 1)^2}\Omega^2}$$

where subscripts b and f denote the backward and forward waves, respectively
[b] Calculated from equation from Soedel [1993]

$$\omega_{mn} = \sqrt{\frac{EH^2}{12(1 - \mu^2)\rho R^2}\frac{1}{R^2}\left[\left(\frac{m\pi R}{L}\right)^2 + n^2\right]\left[\left(\frac{m\pi R}{L}\right)^2 + n^2 - 1\right]}$$

Table 5.2

Comparison of frequency parameter $f = (\omega L^2/H)\sqrt{\rho/E_2}$ for a nonrotating short thick laminated composite (0/90/0°) cylindrical shell ($m = 1$, $n = 2$, $H/R = 0.2$, $E_1 = 40E_2$, $G_{12} = 0.6E_2$, $G_{23} = 0.5E_2$ and $\mu_{12} = 0.25$).

Theory	$L/R = 1$	$L/R = 2$
Present	10.14	18.85
Timarci & Soldatos [1995]		
PAR	9.97	17.16
HYP	9.99	17.16
UNI	9.99	17.16
CST	14.77	20.17
PSDT	10.07	17.77

For the nonrotating cylindrical shells, Tables 5.1 and 5.2 display good agreements between the results available in open literature and those by the present formulations based on the Mindlin shell theory. When the cylindrical shell is thin and long, the agreement achieved as seen in Table 5.1 is especially good for circumferential wave number $n > 3$. Meanwhile, the comparisons are also quite acceptable when the cylindrical shell is thick and short as shown in Table 5.2, in which PAR (parabolic), HYP (hyperbolic) and UNI (uniform) represent the various types of shape functions for the transverse shear deformation profile, and CST (classical shell theory) and PSDT (parabolic shear deformable shell theory) denote different shell theories (see Timarci & Soldatos, 1995). For simply supported rotating cylindrical shells, Tables 5.3 and 5.4 affirm that the present analysis is consistent with available published data, where

Table 5.3

Comparison of frequency parameter $f = \omega R \sqrt{(1 - \mu_{12}^2)\rho/E_2}$ for a rotating long thick cylindrical shell ($\Omega = 0.01$ rad/s, $m = 1$, $H/R = 0.2$, $L/R = 200$, $G_{12} = 0.5E_2$, $\mu_{12} = 0.3$).

	Simplified theory	Soedel [1993]		Present	
n	Soedel [1993]	f_b	f_f	f_b	f_f
1	0.0950806	0.0428968	0.0438266	0.0425013	0.0425106
2	0.209672	0.157745	0.157752	0.210515	0.210523
3	0.513568	0.431536	0.431542	0.486108	0.486114
4	0.937629	0.793428	0.793432	0.847591	0.847595
5	1.22303	1.22303	1.22304	1.27603	1.27603
6	2.14836	1.70367	1.70368	1.75521	1.75522
7	2.93521	2.2221	2.2221	2.2721	2.2721

Table 5.4

Comparison of frequency parameter $f = \omega R \sqrt{(1 - \mu_{12}^2)\rho/E_2}$ for a rotating short thin laminated composite cylindrical shell ($\Omega = 0.1$ rad/s, $m = 1$, $H/R = 0.002$, $E_1 = 2.5E_2$, $G_{12} = G_{13}$, $G_{23} = 0.2E_2$, $\mu_{12} = 0.26$).

L/R	n	Lam & Loy [1994] f_b	Lam & Loy [1994] f_f	Present f_b	Present f_f
1	1	1.061429	1.061140	1.06126	1.06131
	2	0.804214	0.803894	0.804033	0.804084
	3	0.598476	0.598187	0.598325	0.598371
	4	0.450270	0.450021	0.450171	0.45021
	5	0.345363	0.345356	0.345348	0.345392
	6	0.270852	0.270667	0.270988	0.271018
	7	0.217651	0.217489	0.218059	0.218085
5	1	0.248917	0.248917	0.24859	0.024868
	2	0.107436	0.106972	0.107181	0.107254
	3	0.055267	0.054916	0.0551545	0.0552104
	4	0.033945	0.033669	0.0341573	0.0342011
	5	0.025943	0.025718	0.0268394	0.0268753
	6	0.026026	0.025836	0.0278363	0.0278665
	7	0.031089	0.030925	0.0337889	0.033815

comparisons are in general better for thinner shells. Usually the thicker the shell is, the more pronounced will be the shear effects on the natural frequencies. Also, the shorter the cylindrical shell is, the larger will be the effects of boundary condition on the natural frequencies. In addition, when the circumferential wave number n is very small, e.g., $n = 1$, the influence of boundary condition on the frequency characteristics becomes dominant. Thus, the above discussions lead us to the intuitive conclusions that frequency analyses based on the present formulation become less accurate as the shells become very short or too thick. Following careful consideration of the above comparisons and discussions, it can be concluded that the presently developed formulations based on the Mindlin shell theory are accurate and efficient for the dynamic analyses of moderately thick rotating cylindrical shells.

c)   *Frequency characteristics.*

It is well known that as the thickness of cylindrical shells increases, the influence of transverse shear forces becomes more significant on the frequency characteristics of the rotating shell. Therefore, it becomes necessary to understand the differences in frequency characteristics between thick and thin rotating cylindrical shells, and to study the effects of physical and geometrical parameters on the frequency characteristics of thick rotating cylindrical shells. For simplicity in the parametric studies, a rotating equi-thickness laminated composite (0/90/0°) cylindrical shell with the simply supported boundary condition is considered. Material properties are taken as $E_1 = 2.5E_2$, $E_2 = 7.6 \times 10^9$ Pa, $G_{12} = G_{13}$, $G_{23} = 0.2E_2$, $\mu_{12} = 0.26$, $\rho = 1643$ kg/m^3 and the shear correction factor used is $K = 5/6$. The vibrational mode $(m, n) = (1, 1)$ is examined here in the parametric studies although this mode may not necessarily correspond to the fundamental frequency of the rotating cylindrical shell.

Figure 5.1 compares the difference in frequency characteristics between thick and thin rotating cylindrical shells, and shows the variation of frequency parameter $f = \omega R\sqrt{(1 - \mu_{12}^2)\rho/E_2}$ with the rotating speed Ω. As expected, the frequency parameters of both backward and forward waves for the thick rotating shell are always larger than those of a corresponding thin rotating shell. By increasing the rotating speed Ω, the frequency parameters of the backward waves for both thick and thin rotating cylindrical shells increase monotonically while those of forward wave decreases monotonically.

Variations of the frequency parameter $f = \omega R\sqrt{(1 - \mu_{12}^2)\rho/E_2}$ (backward wave) of the rotating laminated composite cylindrical shell with circumferential wave number n

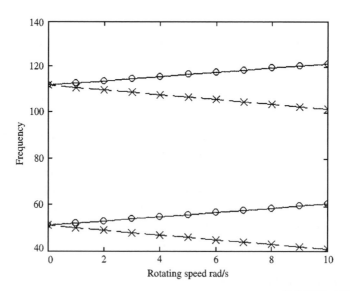

Figure 5.1 Variations of frequency parameter $f = \omega R\sqrt{(1 - \mu_{12}^2)\rho/E_2}$ with the rotating speed for thick ($H/R = 0.2$, above) and thin ($H/R = 0.002$, below) rotating laminated composite cylindrical shells ($L/R = 20$), respectively (solid line for backward wave and dashed line for forward wave).

are illustrated in Fig. 5.2 for a thick shell ($H/R = 0.2$), and in Fig. 5.3 for a thin shell ($H/R = 0.002$). Figure 5.3 shows that the frequency parameter of the thin rotating shell first decreases and then increases with the circumferential wave number. However, as seen in Fig. 5.2, the corresponding frequency parameters of the thick rotating shell increase from the onset of the increase in the circumferential wave number n.

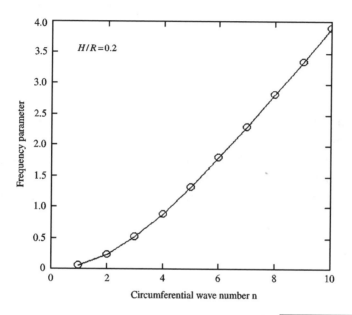

Figure 5.2 Variation of frequency parameter $f = \omega R \sqrt{(1 - \mu_{12}^2)\rho/E_2}$ (backward wave) with the circumferential wave number n for a thick rotating laminated composite cylindrical shell ($m = 1$, $H/R = 0.2$, $L/R = 20$).

Variations of the frequency parameter $f = \omega R \sqrt{(1 - \mu_{12}^2)\rho/E_2}$ (backward wave) of the rotating laminated composite cylindrical shell with different geometric parameter ratios are depicted in Fig. 5.4 for the thickness ratio H/R, and in Fig. 5.5 for the length ratio L/R. Figure 5.4 shows that, with the increase of the thickness ratio H/R, the frequency parameter increases very slowly, and the frequency values for the short rotating shell ($L/R = 5$) are generally larger than those of the long rotating shell ($L/R = 20$). However, Fig. 5.5 shows that the frequency parameter decreases rapidly with the increase of length ratio L/R, and the frequency values for the thick rotating shell ($H/R = 0.2$) are generally larger than those of the corresponding thin rotating shell ($H/R = 0.002$).

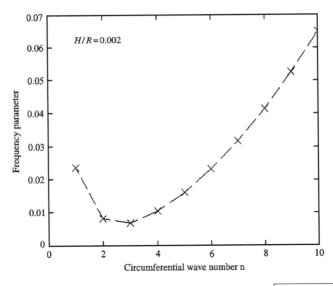

Figure 5.3 Variation of frequency parameter $f = \omega R\sqrt{(1 - \mu_{12}^2)\rho/E_2}$ (backward wave) with the circumferential wave number n for a thin rotating laminated composite cylindrical shell ($m = 1$, $H/R = 0.002$, $L/R = 20$).

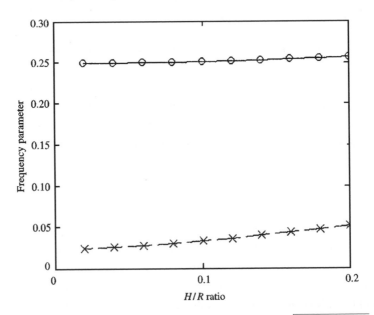

Figure 5.4 Variation of frequency parameter $f = \omega R\sqrt{(1 - \mu_{12}^2)\rho/E_2}$ (backward wave) with the thickness ratio H/R for a rotating laminated composite cylindrical shells ($m = 1$, $n = 1$, $L/R = 5$ (solid line) and 20 (dashed line), respectively).

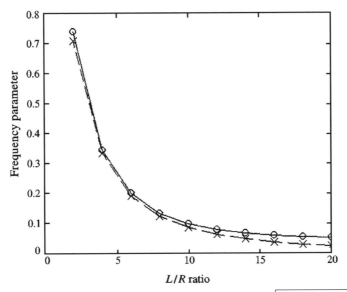

Figure 5.5 Variation of frequency parameter $f = \omega R\sqrt{(1 - \mu_{12}^2)\rho/E_2}$ (backward wave) with the length ratio L/R for a rotating laminated composite cylindrical shells ($m = 1, n = 1, H/R = 0.2$ (solid line) and 0.002 (dashed line), respectively).

5.3   Analysis of vibrational mode by FEM with nonlinear kinematics.

In general, the modal forms in vibrating cylindrical shells are studied and described in the longitudinal and circumferential directions, through the numbers of longitudinal and circumferential waves or the numbers of longitudinal and circumferential nodes. When the shells are sufficiently thin, it is acceptable to describe the vibrational modes by the longitudinal and circumferential wave numbers. For thick cylindrical shells, however, it is difficult based simply on the numbers of longitudinal and circumferential waves to achieve accurate descriptions of the vibrational modes. In order to overcome this, Wang *et al.* [1998] proposed an approach based on the three-dimensional profiles of vibrational mode shapes and successfully classified into eight categories for all the vibrational modes of thick finite-length cylindrical shells. Later, Guo *et al.* [2001] employed this approach for the three-dimensional vibration analysis of thick rotating cylindrical shells.

a)   Classification of three-dimensional modes of thick rotating cylindrical shells.

It is well known that the vibrational modes of rotating and nonrotating thick cylindrical shells are three dimensional. This means that all three displacement components may be associated with each natural frequency. For example, a pure radial excitation will result in responses not only in radial direction, but also in the other two orthogonal directions. Therefore, for an accurate analysis of the vibrational modes of thick rotating cylindrical shell, it is necessary to propose a classification method based on the three-dimensional vibrational modes together with the numbers of longitudinal and circumferential nodes. For simplicity, a thick rotating cylindrical shell with free boundary condition at both ends is taken as an example in the following illustrations. Two integer parameters m and n are redefined here as the number of longitudinal nodes and half of the number of circumferential nodes, respectively. Through numerical simulations and physical experiments, it is verified that all three-dimensional mode shapes in the free vibration of thick cylindrical shells may be classified into the following eight categories:

(1) *Pure radial mode* (see Fig. 5.6) — the vibration is predominantly radial motion. The cylindrical shell retains an almost constant radially deformed cross-sectional shape along its entire length. In addition, the cross-sections remain plane and normal to the shell axis. For this mode, the number of circumferential nodes is $2n$, while the number of longitudinal nodes is always zero. These are important modes because the lowest or fundamental mode of a thick cylindrical shell is a pure radial mode with $n = 2$.

Figure 5.6 Pure radial mode ($m = 0$, $n = 3$).

(2) *Radial shearing mode* (see Fig. 5.7) — for this mode, the cylindrical shell does not possess constant cross-sections, which no longer remain planar. When $m \geq 2$, the generatrices are no longer straight lines, and are therefore not mutually parallel. It becomes clear here that the usual circumferential and longitudinal modes alone are

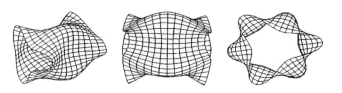

Figure 5.7 Radial shearing mode ($m = 2$, $n = 3$).

not complete mode descriptors because they are the projections of the actual three-dimensional mode shapes on the respective circumferential and longitudinal projective planes.

(3) *Extensional mode* (see Fig. 5.8) — commonly known as breathing modes, the median surface of the shell undergoes uniform strain along each circumference (or cross-section). For these modes, n is zero, and m is the number of longitudinal nodes from the projection of the three-dimensional mode shape on the longitudinal projective plane.

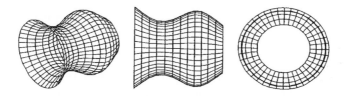

Figure 5.8 Extensional mode ($m = 3$, $n = 0$).

(4) *Circumferential mode* (see Fig. 5.9) — in the circumferential direction, segmental elements can either expand or contract, with the median circumferential length of an expanding segment becoming longer and that of a contracting segment becoming shorter. A nodal radius thus results when there is no net circumferential displacement of a segmental element. For these modes, the number of nodal radii is $2n$, while the number of longitudinal nodes m is dependent upon phase relationships between corresponding nodal radii along the length of the shell.

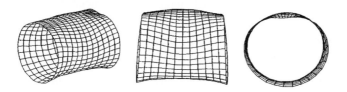

Figure 5.9 Circumferential mode ($m = 1$, $n = 1$).

(5) *Axial bending mode* (see Fig. 5.10) — these modes can be viewed by dividing the circumferential cross-section into several segments, where adjacent segments bend axially in opposing directions. The number of nodal radii is $2n$, and the nodal radial lines exist due to positions along the circumferential cross-section having zero curvature of the transverse plane. For these modes, the number of locations where

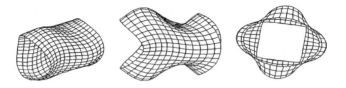

Figure 5.10 Axial bending mode ($m = 0, n = 2$).

the curvature of the deformed cross-section is zero along a given generatrix determines the number of longitudinal modes m.

(6) *Global bending mode* (see Fig. 5.11) — these modes depict the thick cylindrical shell behaving as a simple beam in pure transverse vibration. For these modes, the nodal cross-section is stationary in the transverse direction, and the number of circumferential nodes is $n = 1$. The number of longitudinal nodes m is determined by the projected mode shape of the deformed neutral axis of the cylindrical shell.

Figure 5.11 Global bending mode ($m = 3, n = 1$).

(7) *Torsional mode* (see Fig. 5.12) — these modes depict the thick cylindrical shell behaving as a bar in pure torsional vibration. The number of circumferential nodes n is zero. For these modes, we note a deformed generatrix, but there also exists undeformed nodal transverse planes where the cross-section is stationary in the rotary direction. Thus, the number of longitudinal nodes m is determined by the number of intersections when a deformed generatrix is projected onto the same undeformed generatrix.

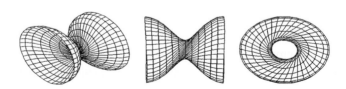

Figure 5.12 Torsional mode ($m = 1, n = 0$).

(8)   *Longitudinal mode* (see Fig. 5.13) — these modes depict the thick cylindrical shell behaving as a rod in pure longitudinal vibration. For these modes, the nodal cross-section is stationary in the axial direction, and the number of circumferential nodes n is zero. The number of longitudinal modes m is determined by the number of locations along the axis of the cylindrical shell whereby we find axially undeformed elements.

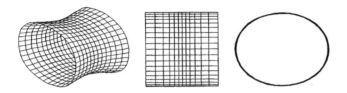

Figure 5.13 Longitudinal mode ($m = 0$, $n = 0$).

If a concept of average displacements $\bar{U}_k$ ($k = 1$, r and t, corresponding to the longitudinal, radial and tangential directions of nodes, respectively) is defined as

$$\bar{U}_k = \frac{100\sqrt{\dfrac{1}{N}\sum\limits_{i=1}^{N} u_{ki}^2}}{\left(\sum\limits_{k}\sqrt{\dfrac{1}{N}\sum\limits_{i=1}^{N} u_{ki}^2}\right)} \qquad (k = 1,\ r,\ t) \tag{5.35}$$

where N is the total number of finite-element nodes of the thick rotating cylindrical shell and u_{ki} is the displacement of node i corresponding to the direction k, several conclusions can then be drawn as follows:

(1) $\bar{U}_1$ of pure radial mode approaches zero,

(2) $\bar{U}_1$ and $\bar{U}_r$ of torsional mode approach zero,

(3) $\bar{U}_t$ of extensional and longitudinal modes approach zero,

(4) $\bar{U}_1$ of axial bending and longitudinal modes is the largest among $\bar{U}_1$, $\bar{U}_r$ and $\bar{U}_t$,

(5) $\bar{U}_r$ of pure radial and radial shearing modes is the largest among $\bar{U}_1$, $\bar{U}_r$ and $\bar{U}_t$,

(6) $\bar{U}_t$ of torsional and circumferential modes is the largest among $\bar{U}_1$, $\bar{U}_r$ and $\bar{U}_t$.

Based on the above numerical experiment and qualitative description, it can be concluded that in general, the present classification approach, based on the definitions of three-dimensional vibrational modes together with the numbers of longitudinal and circumferential nodes, is suitable for describing the three-dimensional modes of thick rotating cylindrical shells. This results in a more accurate analysis of the vibrational modes of thick rotating cylindrical shells ensuring that no mode is missed. It should be noted that the results depicted in Figs. 5.6–5.13 for the eight three-dimensional vibration modes are obtained numerically by FEM computations for a thick rotating cylindrical shell with free boundary condition at both ends. For other boundary conditions, the present classification is also suitable.

b)   *Numerical implementation.*

In order to discuss the qualitative influence of rotation on the three-dimensional vibration modes of a thick rotating cylindrical shell, a finite element technique is implemented using nine-node superparametric curvilinear elements which consider the effects of shear and axial deformations as well as rotatory inertia. In the present numerical implementation based on the moderately thick shell theory, the vibrations of the thick rotating cylindrical shell are assumed to be small around the equilibrium position. The nonlinear plate-shell theory for large deformation is employed to analyze the rotating shell before it reaches the centrifugal induced equilibrium state, upon which a linear approximation is used. The numerical implementation includes the effects of Coriolis acceleration, centrifugal force, initial tension and geometric nonlinearity due to large deflection. The classic nonlinear strain–displacement relations (kinematics) used in the local coordinates x, y, z are given as follows

$$\varepsilon_x = \frac{\partial u}{\partial x} + \frac{1}{2}\left[\left(\frac{\partial u}{\partial x}\right)^2 + \left(\frac{\partial v}{\partial x}\right)^2 + \left(\frac{\partial w}{\partial x}\right)^2\right] \tag{5.36}$$

$$\varepsilon_y = \frac{\partial v}{\partial y} + \frac{1}{2}\left[\left(\frac{\partial u}{\partial y}\right)^2 + \left(\frac{\partial v}{\partial y}\right)^2 + \left(\frac{\partial w}{\partial y}\right)^2\right] \tag{5.37}$$

$$\varepsilon_{xy} = \frac{\partial u}{\partial y} + \frac{\partial v}{\partial x} + \frac{\partial u}{\partial x}\frac{\partial u}{\partial y} + \frac{\partial v}{\partial x}\frac{\partial v}{\partial y} + \frac{\partial w}{\partial x}\frac{\partial w}{\partial y} \tag{5.38}$$

$$\varepsilon_{xz} = \frac{\partial u}{\partial z} + \frac{\partial w}{\partial x} + \frac{\partial u}{\partial y}\frac{\partial u}{\partial z} + \frac{\partial v}{\partial y}\frac{\partial v}{\partial z} \tag{5.39}$$

$$\varepsilon_{yz} = \frac{\partial v}{\partial z} + \frac{\partial w}{\partial y} + \frac{\partial u}{\partial z}\frac{\partial u}{\partial x} + \frac{\partial v}{\partial z}\frac{\partial v}{\partial y}. \tag{5.40}$$

For the simplification of subsequent numerical studies in the following section, free boundary conditions are imposed at both ends of the thick rotating cylindrical shell, while geometric parameters and isotropic material properties are taken as $L = 0.254$ m, $R = 0.09525$ m, $H = 0.0381$ m, $E = 2.07 \times 10^{11}$ N/m^2, $\mu = 0.28$, $\rho = 7.86 \times 10^3$ kg/m^3.

c)   *Influence of rotation on frequencies of various three-dimensional modes.*

In this section, the discussions focus on the influence of rotating velocity on the frequency characteristics of various three-dimensional modes in the vibration of thick rotating free cylindrical shells. The numerically computed frequencies and vibrational modes are tabulated in Table 5.5. It is observed from the table that several similar combinations of m and n exist, such as modes 11 and 12, and modes 14 and 15, which is not possible in stationary shells. For the rotating cylindrical shell, however, two different

Table 5.5

Natural frequencies and vibrational modes of a thick rotating isotropic cylindrical shell with free boundary condition at both ends ($\Omega = 50$ Hz).

Mode	Frequency (Hz)	Displacement ratio $\bar{U}_r : \bar{U}_t : \bar{U}_l$	m	n	Mode type
1	2477	52:26:0	0	2	Pure radial
2	2580	50:25:0	0	2	Pure radial
3	2866	35:19:13	1	2	Radial shearing
4	2958	34:18:12	1	2	Radial shearing
5	6202	24:12:27	1(0)	1	Axial bending
6	6253	24:12:26	1(0)	1	Axial bending
7	6315	0:36:0	1	0	Torsion
8	6991	12:6:4	2	1	Global bending
9	7075	12:6:4	2	1	Global bending
10	8759	13:0:1	2	0	Extensional
11	9502	12:6:19	3(0)	2	Axial bending
12	9526	12:6:19	3(0)	2	Axial bending
13	11,216	16:0:30	2(0)	0(0)	Longitudinal
14	11,726	9:2:4	3	2	Radial shearing
15	11,759	9:2:4	3	2	Radial shearing
16	12,474	6:8:3	1	1	Circumferential
17	12,578	6:8:3	1	1	Circumferential

frequencies associated with the backward and forward waves due to rotation, have the same m and n. Next, variations of the frequencies with the rotating velocities for different three-dimensional modes are illustrated in Figs. 5.14–5.22. In these figures, the ordinate and abscissa represent, respectively, a normalized natural circular frequency ω^* defined as $\omega^* = \omega/\omega_0$ and a normalized rotating speed Ω^* defined as $\Omega^* = \Omega/\omega_0$, where Ω is the rotating speed (Hz), while ω and ω_0 are the frequencies of the rotating shell at speeds Ω and zero (stationary), respectively.

For the pure radial and radial shearing modes, the influences of rotation on the frequency characteristics are shown in Figs. 5.14–5.16 for the thick rotating free cylindrical shell. Figure 5.14 shows the variation of the normalized natural circular frequency ω^* with the normalized rotating speed Ω^* for the pure radial mode ($m = 0$, $n = 2$) and the radial shearing modes ($m = 1$, $n = 2$) and ($m = 2$, $n = 2$). It is observed that the backward-wave frequencies increase and forward-wave frequencies decrease with rotation. For the present case of $n = 2$, as the longitudinal node number m increases from 0 to 2, the backward-wave frequencies decrease and the forward-wave frequencies increase. The frequency difference between backward and forward waves increases

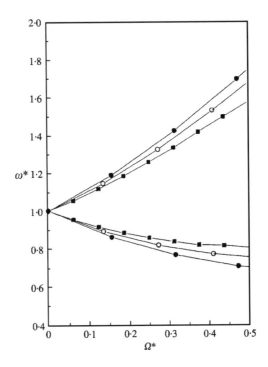

Figure 5.14 Variation of frequency ω^* with rotating speed Ω^* for the pure radial mode ($m = 0$, $n = 2$) ●, and radial shearing modes ($m = 1$, $n = 2$) ○, and ($m = 2$, $n = 2$) ■.

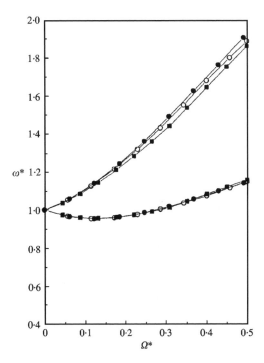

Figure 5.15 Variation of frequency ω^* with
rotating speed Ω^* for the pure radial mode
$(m = 0, n = 3)$ ●, and radial shearing modes
$(m = 1, n = 3)$ ○, and $(m = 2, n = 3)$ ■.

monotonically with rotation. The frequency differences between these three modes are small at lower rotating speeds and subsequently increase with rotation. This trend is also observed in Fig. 5.15 for the pure radial mode $(m = 0, n = 3)$ and the radial shearing modes $(m = 1, n = 3)$ and $(m = 2, n = 3)$, and in Fig. 5.16 for the pure radial mode $(m = 0, n = 4)$ and the radial shearing mode $(m = 1, n = 4)$. In both the latter figures, however, the forward-wave frequencies do not decrease with rotation. Moreover, comparing Figs. 5.14–5.16, it is observed that the frequency difference between the backward and forward waves decreases with the increase of the circumferential node number. At larger circumferential node number, the frequencies of the pure radial and radial shearing modes may be mainly considered as functions of rotating speed.

For the circumferential mode, the variation of the normalized natural circular frequency ω^* with the normalized rotating speed Ω^* is illustrated in Fig. 5.17 for the modes corresponding to $(m = 1, n = 1)$ and $(m = 0, n = 1)$. It is well known that the detection of the circumferential modes is not often possible by the conventional

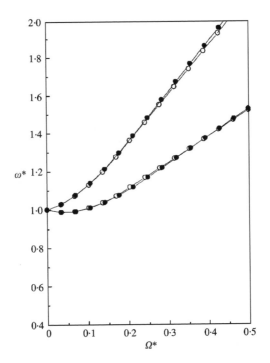

Figure 5.16 Variation of frequency ω^* with rotating speed Ω^* for the pure radial mode $(m = 0, n = 4)$ ●, and the radial shearing mode $(m = 1, n = 4)$ ○.

approach. The frequencies associated with the backward wave are found to increase monotonically while those of the forward wave decrease gradually with rotation.

For the global bending mode, the variation of the normalized natural frequency ω^* with the normalized rotating speed Ω^* is depicted in Fig. 5.18 for the modes corresponding to $(m = 2, n = 1)$ and $(m = 3, n = 1)$. It is observed that, when the rotating speed is small, the frequency variation with the rotating speed may be considered as a linear relation for both the backward and forward waves. However, the variation becomes nonlinear when the rotation increases to larger values. Additionally, there is an obvious influence of the longitudinal node number on the frequency variations of both the backward and forward waves, especially at higher rotating speeds. Similarly, for a given circumferential node number, as the longitudinal node number increases, the backward-wave frequencies decrease and the forward-wave frequencies increase.

For the axial bending mode, the variation of the normalized circular frequency ω^* with the normalized rotating speed Ω^* is shown in Fig. 5.19 for the modes corresponding to $(m = 0, n = 1)$, $(m = 0, n = 2)$ and $(m = 1, n = 1)$. Usually it is difficult to detect

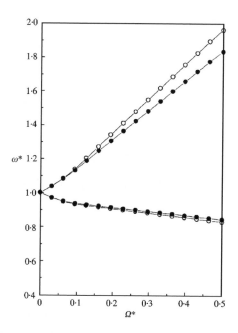

Figure 5.17 Variation of frequency ω^* with rotating speed Ω^* for the circumferential modes ($m = 1, n = 1$) ●, and ($m = 0, n = 1$) ○.

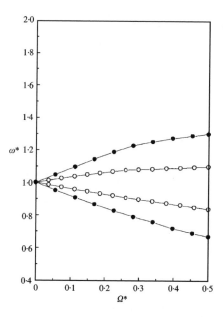

Figure 5.18 Variation of frequency ω^* with rotating speed Ω^* for the global bending modes ($m = 2, n = 1$) ●, and ($m = 3, n = 1$) ○.

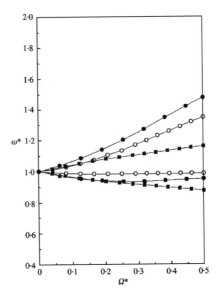

Figure 5.19 Variation of frequency ω^* with rotating speed Ω^*
for the axial bending modes $(m = 0, n = 1)$ ●,
$(m = 0, n = 2)$ ○, and $(m = 1, n = 1)$ ■.

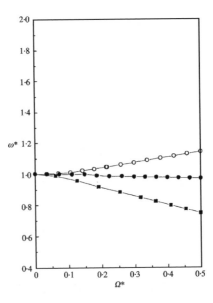

Figure 5.20 Variation of frequency ω^* with rotating speed Ω^* for the
extensional mode $(m = 3, n = 0)$ ●, longitudinal mode $(m = 0, n = 0)$ ○, and
torsional mode $(m = 1, n = 0)$ ■.

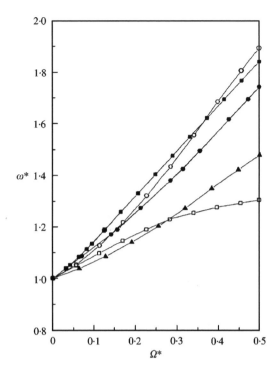

Figure 5.21 Variation of backward-wave frequency ω^* with rotating speed Ω^* for the pure radial mode ($m = 0, n = 2$) ●, radial shearing mode ($m = 1, n = 3$) ○, circumferential mode ($m = 0, n = 1$) ■, global bending mode ($m = 2, n = 1$) □, and axial bending mode ($m = 0, n = 1$) ▲.

the axial bending modes by the conventional approach. Once again it is observed here that, as the rotating speed increases, the frequencies of the backward waves increase monotonically, and the frequencies associated with modes ($m = 0, n = 1$) and ($m = 0, n = 2$) increase more steeply than those of mode ($m = 1, n = 1$). For the forward waves, the frequencies associated with mode ($m = 1, n = 1$) decrease with rotation, while those of modes ($m = 0, n = 1$) and ($m = 0, n = 2$) first decrease and then subsequently remain almost constant. It is also seen that the influences of both numbers of the longitudinal and circumferential nodes are significant on the frequency characteristics of both the backward and forward waves.

For a quantitative comparison of three different mode forms — extensional, longitudinal and torsional, the variation of the normalized frequency ω^* with the normalized rotating speed Ω^* is shown in Fig. 5.20 for the respective modes (($m = 3, n = 0$) extensional mode, ($m = 0, n = 0$) longitudinal mode and ($m = 1, n = 0$) torsional

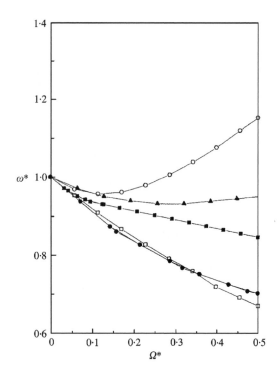

Figure 5.22 Variation of forward-wave frequency ω^* with rotating speed Ω^* for the pure radial mode ($m = 0$, $n = 2$) ●, radial shearing mode ($m = 1$, $n = 3$) ○, circumferential mode ($m = 0$, $n = 1$) ■, global bending mode ($m = 2$, $n = 1$) □, and axial bending mode ($m = 0$, $n = 1$) ▲.

mode). It is obvious here that, when the circumferential node number is zero, there are three possible distinct vibrational mode forms, namely the extensional, longitudinal and torsional modes. However, only one frequency exists for each mode at a given rotating speed. As such, only one frequency curve exists for each mode as shown in Fig. 5.20. For the extensional mode ($m = 3$, $n = 0$), there is only slight frequency variation with rotation. For the longitudinal mode ($m = 0$, $n = 0$), the frequency increases gradually with rotation. For the torsional mode ($m = 1$, $n = 0$), the frequency decreases with rotation. In the FEM, we note here for the present rotating-shell problem that the global stiffness matrix consists of three components, i.e., the stiffness terms due to shell geometry and material properties, those due to pre-stress when the shell rotates, which increases the total stiffness, and those due to variation of centrifugal force, which reduces the total stiffness. Thus, it is very difficult to predict the trends of the frequencies without actually carrying out the numerical computations.

Finally, to carry out a quantitative comparison on the other five modes, namely the pure radial mode ($m = 0, n = 2$), radial shearing mode ($m = 1, n = 3$), circumferential mode ($m = 0, n = 1$), global bending mode ($m = 2, n = 1$) and the axial bending mode ($m = 0, n = 1$), the variations of the normalized natural circular frequency ω^* with the normalized rotating speed Ω^* are illustrated in Fig. 5.21 for backward waves and in Fig. 5.22 for forward waves. One can observe that the frequency differences of the different modes are small when the rotating speed is small, and these differences will significantly increase as the rotation increases. In addition, it is noted from Fig. 5.22 that the forward-wave frequencies of the radial shearing mode ($m = 1, n = 3$) increase with the increase of rotating speed, while the rest of the forward-wave frequencies associated with the other four modes display general decreasing trends.

Chapter 6
Critical Speed and Dynamic Stability of Thin Rotating Isotropic Cylindrical Shells

6.1   Introduction.

The critical speed analysis of shaft–disc systems has been extensively investigated over the last hundred years whereas similar analyses on cylindrical shells or drum-like rotor structures have been few and far between. The first published work on this problem was by Bryan [1890] where a rotating thin ring was considered and it was here that the phenomenon of traveling modes was discovered. The effects of Coriolis forces were first investigated by DiTaranto & Lessen [1964] for an infinitely long isotropic cylindrical shell. It was concluded that the Coriolis effects have significant influence on the natural frequencies and should always be considered in such analyses. Srinivasan & Lauterbach [1971] then combined both the effects of the Coriolis forces and traveling modes in the study of rotating isotropic cylindrical shells. Zohar & Aboudi [1973] later applied both these effects in their investigations on finite length rotating shells. However, these early works concentrated mainly on frequency analysis.

The first critical speed results for rotating shells were obtained experimentally by Zinberg & Symonds [1970]. The results also proved the advantages of using shells made of orthotropic materials over aluminum alloy shells. A finite element approach was used by dos Reis, Goldman, & Verstrate [1987] to obtain the critical speeds in the evaluation of the shell of Zinberg & Symonds [1970]. More recently, a simplified theory for analyzing the first critical speed of a composite cylindrical shell was performed by Kim & Bert [1993]. Results obtained using various shell theories were compared.

Due to its many complexities, a literature search will show that studies on the dynamic stability of rotating cylindrical shells have received even less attention than the above-mentioned critical speed analyses. Such studies would be interesting and important as it would shed light on the qualitative and quantitative effects of centrifugal and Coriolis forces on the instability regions.

Thus in this chapter, a theoretical basis is presented to study the critical speed and dynamic stability of rotating cylindrical shells under axial loading. For the critical speed analysis, we will use Donnell's theory for thin shells together with consideration

229

of the initial hoop tension and centrifugal and Coriolis forces. For the dynamic stability analysis of thin rotating cylindrical shells under combined static and periodic axial forces, we will once again employ the Donnell's theory for thin shells with consideration of initial hoop tension and the centrifugal and Coriolis forces. In this dynamic stability analysis, a normal-mode expansion yields a system of Mathieu–Hill equations. The parametric resonance response is analyzed based on Bolotin's method. The two above-mentioned problems are formulated to allow the cylindrical shell assume any boundary conditions, but for simplicity, results are presented only for the case of simply supported shells as displacement fields which satisfy these boundary conditions and can be easily expressed in terms of the products of sine and cosine functions. Numerical results for the critical speeds will be presented for the relevant modes, while those for the instability regions will be presented for different transverse modes at various rotational speeds.

6.2   Theoretical development: axially loaded rotating shells.

The cylindrical shell as shown in Fig. 3.1 is assumed to be a thin, uniform shell of length L, thickness h, radius R and rotating about the x-axis at constant angular velocity Ω. The elastic modulus is denoted by E, mass density by ρ and Poisson's ratio by ν. The x-axis is taken along a generator, the circumferential arc length subtends an angle θ, and the z-axis is directed radially inwards.

For critical speed analysis, the extensional axial load per unit length is constant and denoted by N_o which in nondimensional form is given by

$$\eta_o = \frac{N_o(1 - \nu^2)}{Eh} \tag{6.1}$$

For this analysis, Donnell's theory for a thin-walled cylindrical shell is assumed to lead to the following three equations of motion:

$$R^2 \frac{\partial^2 u}{\partial x^2} + \frac{1}{2}(1 - \nu)\frac{\partial^2 u}{\partial \theta^2} + \frac{R}{2}(1 + \nu)\frac{\partial^2 v}{\partial x \partial \theta} + \nu R \frac{\partial w}{\partial x}$$

$$+ \frac{\gamma}{\rho h}\tilde{N}_\theta \left(\frac{1}{R^2}\frac{\partial^2 u}{\partial \theta^2} - \frac{1}{R}\frac{\partial w}{\partial x} \right) = \gamma \frac{\partial^2 u}{\partial t^2} \tag{6.2a}$$

$$\frac{R}{2}(1+v)\frac{\partial^2 u}{\partial x \partial \theta} + \frac{R^2}{2}(1-v)\frac{\partial^2 v}{\partial x^2} + \frac{\partial^2 v}{\partial \theta^2} + \frac{\partial w}{\partial \theta} + \frac{\gamma}{\rho h}\tilde{N}_\theta \frac{1}{R}\frac{\partial^2 u}{\partial x \partial \theta}$$

$$= \gamma\left(\frac{\partial^2 v}{\partial t^2} + 2\Omega\frac{\partial w}{\partial t} - \Omega^2 v\right)$$

(6.2b)

$$-vR\frac{\partial u}{\partial x} - \frac{\partial v}{\partial \theta} - w - k\left(R^4\frac{\partial^4 w}{\partial x^4} + 2R^2\frac{\partial^4 w}{\partial x^2 \partial \theta^2} + \frac{\partial^4 w}{\partial \theta^4}\right)$$

$$+ \frac{\gamma}{\rho h}\tilde{N}_\theta\left(\frac{1}{R^2}\frac{\partial^2 w}{\partial \theta^2} - \frac{1}{R^2}\frac{\partial v}{\partial \theta}\right) + R^2\frac{\partial}{\partial x}\left(\eta_o\frac{\partial w}{\partial x}\right) = \gamma\left(\frac{\partial^2 w}{\partial t^2} - 2\Omega\frac{\partial v}{\partial t} - \Omega^2 w\right)$$

(6.2c)

where u, v and w are the components of displacement for an element of the shell wall. The initial hoop tension due to the centrifugal force is defined as

$$\tilde{N}_\theta = \rho h \Omega^2 R^2$$

(6.3)

and

$$\gamma = \frac{\rho R^2(1-v^2)}{E}$$

(6.4)

$$k = \frac{h^2}{12R^2}$$

(6.5)

For the dynamic stability analysis, the extensional axial load per unit length is now periodic (harmonic) and given in nondimensional form as

$$\eta_a = \eta_o + \eta_s \cos Pt$$

(6.6)

where P is the frequency of excitation in radians per unit time. η_a, η_o and η_s are the nondimensionalized load parameters defined as

$$\eta_a = \frac{N_a(1-v^2)}{Eh}, \qquad \eta_o = \frac{N_o(1-v^2)}{Eh}, \qquad \eta_s = \frac{N_s(1-v^2)}{Eh}$$

(6.7)

where N_a is the axial loading per unit length (N/m), with N_o being the constant component and N_s the oscillatory component.

Thus, the equations of motion for the dynamic stability analysis of a rotating cylindrical shell under combined static and periodic loading follow those given in Eq. (6.2a–c) except that in the third equation, η_o should be replaced by η_a.

6.3   Numerical implementation.

Having described the equations of motions for the critical speed and dynamic stability analyses of axially loaded rotating cylindrical shells, the two following sub-sections will detail the numerical procedures by which the critical speeds and the boundaries of the instability regions are obtained.

a)   *Critical speed analysis.*

As mentioned earlier, for simplicity, we shall assume the shell to be simply supported. In this case, there exists a solution for the three equations described in Eq. (6.2a–c) in the form

$$u = A \cos\frac{m\pi x}{L}\cos(n\theta + \omega t) \tag{6.8}$$

$$v = B \sin\frac{m\pi x}{L}\sin(n\theta + \omega t) \tag{6.9}$$

$$w = C \sin\frac{m\pi x}{L}\cos(n\theta + \omega t) \tag{6.10}$$

where n represents the number of circumferential waves and m, the number of axial half-waves in the corresponding standing wave pattern.

The equations of motion, Eq. (6.2a–c), are solved using an eigenfunction expansion in terms of the normal modes of the free vibrations of a cylindrical shell. Substitution of Eqs. (6.8)–(6.10) into Eq. (6.2a–c) yields a set of three coupled homogenous equations

$$\begin{bmatrix} C_{11} & C_{12} & C_{13} \\ C_{21} & C_{22} & C_{23} \\ C_{31} & C_{32} & C_{23} \end{bmatrix} \begin{Bmatrix} A \\ B \\ C \end{Bmatrix} = \begin{Bmatrix} 0 \\ 0 \\ 0 \end{Bmatrix} \tag{6.11}$$

where the coefficients $C_{ij}(i, j = 1, 2, 3)$ are given by

$$C_{11} = R^2\lambda^2 + \frac{1}{2}(1-v)n^2 + \frac{1}{R^2}\frac{\gamma}{\rho h}\tilde{N}_\theta n^2 - \gamma\omega^2 + R^2\lambda^2\eta_o$$

$$C_{12} = -\frac{R}{2}(1+v)\lambda n$$

$$C_{13} = -Rv\lambda + \frac{1}{R}\frac{\gamma}{\rho h}\tilde{N}_\theta\lambda$$

$$C_{21} = -\frac{R}{2}(1+v)\lambda n - \frac{1}{R}\frac{\gamma}{\rho h}\tilde{N}_\theta\lambda n$$

$$C_{22} = \frac{R^2}{2}(1-v)\lambda^2 + n^2 - \gamma(\omega^2 + \Omega^2) + R^2\lambda^2\eta_o \tag{6.12}$$

$$C_{23} = n - 2\gamma\omega\Omega$$

$$C_{31} = -Rv\lambda$$

$$C_{32} = n + \frac{1}{R^2}\frac{\gamma}{\rho h}\tilde{N}_\theta n - 2\gamma\omega\Omega$$

$$C_{33} = R^4 k\lambda^4 + 2R^2 k\lambda^2 n^2 + kn^4 + 1 + \frac{1}{R^2}\frac{\gamma}{\rho h}\tilde{N}_\theta n^2 - \gamma(\omega^2 + \Omega^2) + R^2\lambda^2\eta_o$$

and

$$\lambda = \frac{m\pi}{L} \tag{6.13}$$

Imposing non-trivial conditions on Eq. (6.11), the characteristic frequency equation is obtained by equating the determinant of the characteristic matrix in Eq. (6.11) to zero

$$\begin{vmatrix} C_{11} & C_{12} & C_{13} \\ C_{21} & C_{22} & C_{23} \\ C_{31} & C_{32} & C_{23} \end{vmatrix} = 0 \tag{6.14}$$

Expanding Eq. (6.14), a polynomial of the following form can be obtained

$$\beta_1\omega_{mn}^6 + \beta_2\omega_{mn}^4 + \beta_3\omega_{mn}^3 + \beta_4\omega_{mn}^2 + \beta_5\omega_{mn} + \beta_6 = 0 \tag{6.15}$$

and the six roots for each m and n of Eq. (6.15) can be solved using the Newton–Raphson procedure.

The critical speed phenomena is clearly shown in Fig. 6.1. In the figure, Y_{mn} and X denote the normalized natural frequency and rotational speed, respectively, normalized with respect to the non-rotational natural frequency ω_{mno}, i.e.,

$$Y_{mn} = \frac{\omega_{mn}}{\omega_{mno}} \tag{6.16}$$

$$X = \frac{\Omega}{\omega_{mno}} \tag{6.17}$$

From Fig. 6.1, the bifurcations of the natural frequencies for the transverse modes of $(m,n) = (1,1), (1,2), (1,3)$ and $(1,4)$ for a rotating cylindrical shell of geometric properties $L/R = 6$ and $R/h = 50$, and Poisson's ratio $v = 0.3$, are presented. For this illustrative

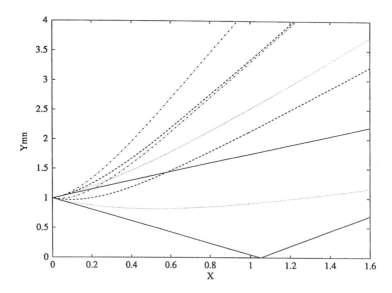

Figure 6.1 Bifurcations of natural frequencies of a rotating cylindrical shell with geometric properties $L/R = 6$ and $R/h = 50$, and $\eta_o = 0$. — $(m, n) = (1, 1)$; ⋯⋯ $(m, n) = (1, 2)$; --- $(m, n) = (1, 3)$; --- $(m, n) = (1, 4)$.

case, there is no axial loading and the results obtained using the present formulation compares well with the results obtained by Huang & Hsu [1990].

Due to the Coriolis acceleration, the nodal lines are neither stationary nor rotating at the same angular velocity as the shell. It can be shown that each mode rotates at its own speed given by $\Omega - \omega_{mno}/n$ with respect to the stationary coordinates. The lower branch corresponds to the forward whirl and the upper branch corresponds to the backward whirl. The critical speed of the rotating shell which occurs for mode (1,1) here corresponds to the rotational speed of the shell at which the forward mode intersects the abscissa. At this intersection, possible unstable phenomenon exists as the forward mode becomes standing with respect to the traveling θ coordinate and is ready to switch to a backward mode. At this critical speed, any residual unbalance will synchronize with the rotation and magnify the whirling amplitude.

b)   *Dynamic stability analysis.*

Here again, for simplicity, we shall assume the shell to be simply supported. Therefore, the solution will be of the form given in Eqs. (6.8)–(6.10). Now for every combination of m and n, there are six distinct natural frequencies. It has been concluded by Huang & Hsu [1990] that in most engineering applications, the transverse modes predominate such that the contribution of in-plane modes, i.e., ω_{mnj}, $j = 3, 4, 5, 6$ can be neglected. Thus, Eqs. (6.8)–(6.10) can be expanded and simplified in terms of two generalized coordinates

$$u_{mnj} = \sum_{j=1}^{2} \sum_{m=1}^{\infty} \sum_{n=1}^{\infty} A_{mnj}[p_{mnj}(t)\cos(n\theta) - q_{mnj}(t)\sin(n\theta)]\cos(\lambda_m x) \tag{6.18}$$

$$v_{mnj} = \sum_{j=1}^{2} \sum_{m=1}^{\infty} \sum_{n=1}^{\infty} B_{mnj}[p_{mnj}(t)\sin(n\theta) + q_{mnj}(t)\cos(n\theta)]\sin(\lambda_m x) \tag{6.19}$$

$$w_{mnj} = \sum_{j=1}^{2} \sum_{m=1}^{\infty} \sum_{n=1}^{\infty} C_{mnj}[p_{mnj}(t)\cos(n\theta) - q_{mnj}(t)\sin(n\theta)]\sin(\lambda_m x) \tag{6.20}$$

where $p_{mnj}(t)$ and $q_{mnj}(t)$ are the generalized coordinates.

Upon substitution of Eqs. (6.18)–(6.20) into Eq. (6.2a–c), we make use of the orthogonality condition by multiplying Eq. (6.2a) separately by $\Gamma_{rsi} \cos \lambda_r x \cos s\theta$ and $\Gamma_{rsi} \cos \lambda_r x \sin s\theta$, Eq. (6.2b) separately by $\beta_{rsi} \sin \lambda_r x \sin s\theta$ and $\beta_{rsi} \sin \lambda_r x \cos s\theta$

and Eq. (6.2c) separately by $\sin \lambda_r x \cos s\theta$ and $\sin \lambda_r x \sin s\theta$ where

$$\Gamma_{mnj} = \frac{A_{mnj}}{C_{mnj}}, \qquad \beta_{mnj} = \frac{B_{mnj}}{C_{mnj}} \tag{6.21}$$

We then add the six resulting equations and integrate over the surface of the shell. This yields the following set of equations:

$$\mathbf{M}^*\ddot{\mathbf{f}} + \mathbf{G}^*\dot{\mathbf{f}} + [\mathbf{K}^* - \cos Pt\mathbf{Q}^*]\mathbf{f} = 0 \tag{6.22}$$

where the matrices $\mathbf{M}^*$, $\mathbf{G}^*$, $\mathbf{K}^*$ and $\mathbf{Q}^*$ are given by

$$\mathbf{M}^* = \begin{bmatrix} \mathbf{M}_{IJ} & 0 \\ 0 & \mathbf{M}_{IJ} \end{bmatrix} \quad \mathbf{K}^* = \begin{bmatrix} \mathbf{K}_{IJ} & 0 \\ 0 & \mathbf{K}_{IJ} \end{bmatrix} \tag{6.23}$$

$$\mathbf{G}^* = \begin{bmatrix} 0 & -\mathbf{G}_{IJ} \\ \mathbf{G}_{IJ} & 0 \end{bmatrix} \quad \mathbf{Q}^* = \begin{bmatrix} \mathbf{Q}_{IJ} & 0 \\ 0 & \mathbf{Q}_{IJ} \end{bmatrix} \tag{6.24}$$

and $\ddot{\mathbf{f}}$, $\dot{\mathbf{f}}$ and $\mathbf{f}$ are column vectors given by

$$\ddot{\mathbf{f}} = \left\{ \begin{array}{c} \ddot{\mathbf{p}}_J \\ \ddot{\mathbf{q}}_J \end{array} \right\}, \qquad \dot{\mathbf{f}} = \left\{ \begin{array}{c} \dot{\mathbf{p}}_J \\ \dot{\mathbf{q}}_J \end{array} \right\}, \qquad \mathbf{f} = \left\{ \begin{array}{c} \mathbf{p}_J \\ \mathbf{q}_J \end{array} \right\} \tag{6.25}$$

The subscripts r, s, i, m, n, j, I and J have the following ranges

$$r = 1, 2, 3, 4, ..., N, \qquad s = 1, 2, 3, 4, ..., N, \qquad i = 1, 2,$$

$$m = 1, 2, 3, 4, ..., N, \qquad n = 1, 2, 3, 4, ..., N, \qquad j = 1, 2, \tag{6.26}$$

$$I = 1, 2, 3, 4, ..., (N \times N \times 2), \qquad J = 1, 2, 3, 4, ..., (N \times N \times 2)$$

and are arranged as

$$
\begin{aligned}
&I = 1, \quad r = 1 \quad s = 1 \quad i = 1 \\
&I = 2, \quad r = 1 \quad s = 1 \quad i = 2 \\
&I = 3, \quad r = 1 \quad s = 2 \quad i = 1 \\
&I = 4, \quad r = 1 \quad s = 2 \quad i = 2
\end{aligned}
$$

$$\vdots$$

$$
\begin{aligned}
&I = 2N - 1, \quad r = 1 \quad s = N \quad i = 1 \\
&I = 2N, \qquad\;\; r = 1 \quad s = N \quad i = 2 \\
&I = 2N + 1, \quad r = 2 \quad s = 1 \quad i = 1 \\
&I = 2N + 2, \quad r = 2 \quad s = 1 \quad i = 2
\end{aligned}
$$

(6.27)

$$\vdots$$

$$
\begin{aligned}
&I = 2N^2 - 1, \quad r = N \quad s = N \quad i = 1 \\
&I = 2N^2, \qquad\;\; r = N \quad s = N \quad i = 2
\end{aligned}
$$

The co-relations between the subscripts J, m, n and j follow those of I, r, s and i, respectively, and will, therefore, not be explicitly expressed here.

The matrices $\mathbf{M}_{IJ}$, $\mathbf{K}_{IJ}$, $\mathbf{G}_{IJ}$ and $\mathbf{Q}_{IJ}$ are given as

$$
\mathbf{M}_{IJ} = \begin{cases} \gamma(\pi L/2)(1 + \beta_I \beta_J + \Gamma_I \Gamma_J) & \text{if } I = J \\ 0 & \text{if } I \neq J \end{cases}
\tag{6.28}
$$

$$
\mathbf{K}_{IJ} = \begin{cases} (\pi L/2)\mathbf{K}^* & \text{if } I = J \\ 0 & \text{if } I \neq J \end{cases}
\tag{6.29}
$$

$$
\mathbf{G}_{IJ} = \begin{cases} (\pi L/2)(2\gamma\Omega)(\beta_I + \beta_J) & \text{if } I = J \\ 0 & \text{if } I \neq J \end{cases}
\tag{6.30}
$$

$$
\mathbf{Q}_{IJ} = \begin{cases} -(\pi L/2)(R^2 \lambda_m \lambda_r \eta_s) & \text{if } I = J \\ 0 & \text{if } I \neq J \end{cases}
\tag{6.31}
$$

where

$$\mathbf{K}^* = \Gamma_I \Gamma_J \left[(R\lambda_m)^2 + \frac{(1-\nu)}{2} n^2 + \frac{\tilde{N}_\theta}{\rho h} \frac{\gamma}{R^2} n^2 \right] - \Gamma_I \beta_J \left[\frac{1+\nu}{2} R\lambda_m n \right]$$

$$- \Gamma_I \left[\nu R\lambda_m - \frac{\tilde{N}_\theta}{\rho h} \frac{\gamma}{R} \lambda_m \right] - \beta_I \Gamma_J \left[\frac{1+\nu}{2} R\lambda_m n + \frac{\tilde{N}_\theta}{\rho h} \frac{\gamma}{R} \lambda_m n \right]$$

$$+ \beta_I \beta_J \left[\frac{(1-\nu)}{2} (R\lambda_m)^2 + n^2 - \gamma\Omega^2 \right] - \beta_I[-n] - \Gamma_J[\nu R\lambda_m]$$

$$- \beta_J \left[-n - \frac{\tilde{N}_\theta}{\rho h} \frac{\gamma}{R^2} n \right]$$

$$+ \left[k((R\lambda_m)^2 + n^2)^2 + 1 + \frac{\tilde{N}_\theta}{\rho h} \frac{\gamma}{R^2} n^2 - \gamma\Omega^2 + \eta_0 (R\lambda_m)^2 \right] \tag{6.32}$$

Equation (6.22) is in the form of a second-order differential equation with periodic coefficients of the Mathieu–Hill type. Using the method presented by Bolotin [1964], the regions of unstable solutions are separated by periodic solutions having period T and $2T$ with $T = 2\pi/P$. The solutions with period $2T$ are of greater practical importance as the widths of these unstable regions are usually larger than those associated with solutions having period T. As a first approximation, the periodic solutions with period $2T$ can be sought in the form

$$\mathbf{f} = \mathbf{a} \sin\frac{Pt}{2} + \mathbf{b} \cos\frac{Pt}{2} \tag{6.33}$$

where $\mathbf{a}$ and $\mathbf{b}$ are arbitrary vectors.

Substituting Eq. (6.33) into Eq. (6.22) and equating the coefficients of $\sin(Pt/2)$ and $\cos(Pt/2)$ terms, a set of linear homogeneous algebraic equations in terms of a and b can be obtained. The conditions for non-trivial solutions are given by

$$\det \begin{bmatrix} \mathbf{K}^* - \frac{1}{2}\mathbf{Q}^* - \frac{1}{4}P^2\mathbf{M}^* & -\frac{1}{2}P\mathbf{G}^* \\ \frac{1}{2}P\mathbf{G}^* & \mathbf{K}^* + \frac{1}{2}\mathbf{Q}^* - \frac{1}{4}P^2\mathbf{M}^* \end{bmatrix} = 0 \tag{6.34}$$

Equation (6.34) is the equation of boundary frequencies and can be used to calculate the boundaries of the instability regions.

6.4   Critical speeds and instability regions.

Having described the procedure for obtaining the critical speeds of rotating cylindrical shells under constant axial loading, and that for obtaining the boundaries of the instability regions of rotating cylindrical shells under combined static and periodic (harmonic) loading, numerical results will now be generated in the two following sub-sections to study the major characteristics of critical speed and dynamic stability in axially loaded rotating cylindrical shells, respectively.

a)   Influence of axial loading on critical speeds.

In this study, the effects of constant axial loading on the natural frequencies and the critical speeds of rotating shells are examined. The results presented in Figs. 6.2–6.6 are for a simply supported cylindrical shell with Poisson's ratio $v = 0.3$ and geometric properties $L/R = 20$ and $R/h = 50$. The modes of interest here are the transverse modes as they correspond to the two lowest natural frequencies. The two higher axial and circumferential modes are not presented. The results presented exclude those for circumferential wave

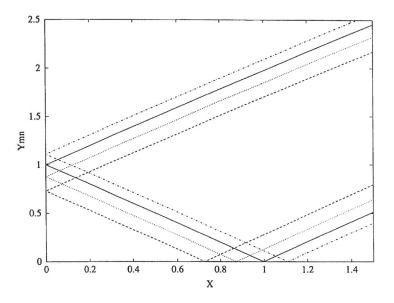

Figure 6.2 Bifurcations of natural frequencies for mode $(m, n) = (1, 1)$ of a rotating cylindrical shell with geometric properties $L/R = 20$ and $R/h = 50$. --- $\eta_o = 0.3\eta_{cri}$; — $\eta_o = 0$; ⋯⋯ $\eta_o = -0.3\eta_{cri}$; --- $\eta_o = -0.6\eta_{cri}$.

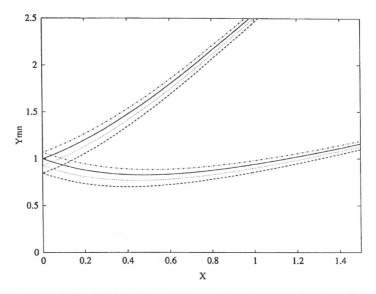

Figure 6.3 Bifurcations of natural frequencies for mode $(m, n) = (1, 2)$ of a rotating cylindrical shell with geometric properties $L/R = 20$ and $R/h = 50$. --- $\eta_{\mathrm{o}} = 0.3\eta_{\mathrm{cri}}$; — $\eta_{\mathrm{o}} = 0$; ······ $\eta_{\mathrm{o}} = -0.3\eta_{\mathrm{cri}}$; --- $\eta_{\mathrm{o}} = -0.6\eta_{\mathrm{cri}}$.

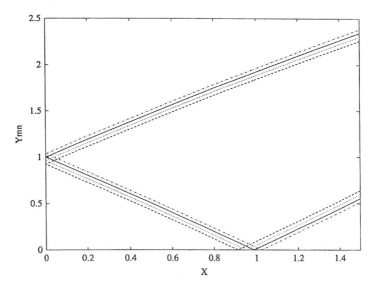

Figure 6.4 Bifurcations of natural frequencies for mode $(m, n) = (2, 1)$ of a rotating cylindrical shell with geometric properties $L/R = 20$ and $R/h = 50$. --- $\eta_{\mathrm{o}} = 0.3\eta_{\mathrm{cri}}$; — $\eta_{\mathrm{o}} = 0$; ······ $\eta_{\mathrm{o}} = -0.3\eta_{\mathrm{cri}}$; --- $\eta_{\mathrm{o}} = -0.6\eta_{\mathrm{cri}}$.

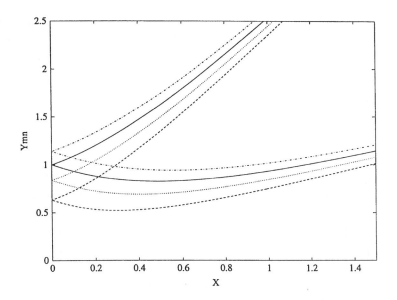

Figure 6.5 Bifurcations of natural frequencies for mode $(m, n) = (2, 2)$ of a rotating cylindrical shell with geometric properties $L/R = 20$ and $R/h = 50$. --- $\eta_o = 0.3\eta_{cri}$; — $\eta_o = 0$; ······ $\eta_o = -0.3\eta_{cri}$; --- $\eta_o = 0.6\eta_{cri}$.

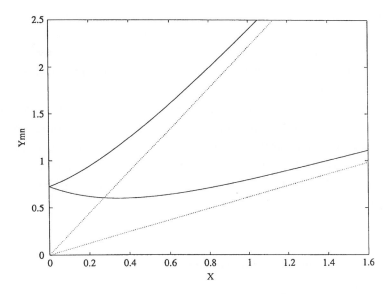

Figure 6.6 Bifurcations of natural frequencies of a rotating cylindrical shell with geometric properties $L/R = 20$ and $R/h = 50$, and $\eta_o = -\eta_{cri}$. — $(m, n) = (1, 2)$; ······ $(m, n) = (2, 2)$.

number $n > 4$ due to the limitation of Donnell's equations to the higher modes for short to moderate length shells. Figures 6.2–6.5 show the results for different loadings for the transverse modes of modes $(m,n) = (1,1)$, $(1,2)$, $(2,1)$, and $(2,2)$, respectively.

It is important to note here that the axial loading, if compressive, must be a fraction of the static critical buckling load, η_{cri}. The governing buckling differential equations can be obtained by neglecting the terms involving $\tilde{N}_\theta$, Ω and t in Eq. (6.2a–c). The buckling loads can be easily obtained by computing the eigenvalues of the resulting characteristic matrix.

For the shell used in Figs. 6.2–6.5, the static critical buckling load, η_{cri}, corresponds to the transverse mode of mode $(2,2)$. The corresponding buckling loads for modes $(1,1)$, $(1,2)$ and $(2,1)$ are 1.287, 2.12, and 4.06 η_{cri}, respectively.

From Figs. 6.2–6.5, it is observed that modes $(1,1)$ and $(2,1)$ exhibit critical speed phenomena while modes $(1,2)$ and $(2,2)$ do not. It is also observed that the response to tensile loading is generally predictable for all the modes with an upward shift for all the branches. This can be expected as increased tensile loading causes the shell to become stiffer. Also, the critical speeds for modes $(1,1)$ and $(2,1)$ are raised due to tensile loading. For compressive loadings, the expected downward shifts for all the branches are also observed. Correspondingly, the critical speeds for modes $(1,1)$ and $(2,1)$ are lowered due to compressive loading.

It is also seen here that although axial loadings can alter the magnitudes of the critical speeds significantly, it cannot induce critical speed phenomena in modes that do not otherwise exhibit this phenomena when not subjected to any axial loading. This is illustrated in Fig. 6.6 where modes $(1,2)$ and $(2,2)$ are given the maximum allowable compressive loading, $-\eta_{cri}$, but the lower branches or the forward modes do not intersect the abscissa.

Finally, it is observed from Figs. 6.2–6.5 that different modes show different levels of sensitivity to the axial loadings. It can be deduced that the level of sensitivity of a particular mode is proportionally dependent to the magnitude of the buckling load of that mode. For example, mode $(2,1)$ which has the highest buckling load is observed to be the least sensitive while mode $(2,2)$ whose buckling load is the critical buckling load, η_{cri}, is observed to be the most sensitive.

b)   ***Parametric studies on dynamic stability.***

The dynamic instability regions for the first-order parametric resonances of a rotating cylindrical shell under combined static and periodic axial loads are presented in Tables 6.1–6.4 and Figs. 6.8–6.11. The nondimensional excitation frequency parameter

Table 6.1

Unstable regions for the transverse modes of a simply supported isotropic rotating cylindrical shell of $v = 0.3$ and geometric properties $L/R = 2$ and $R/h = 100$ and subjected to extensional (tensile) loading of $\eta_o = 0.1\eta_{cri}$.

$\bar{\Omega}$	p_1	p_2	$\Theta(\times 10^{-3})$
Mode (1,1)			
0	1.147510143	1.147510143	1.181011
$0.1\bar{\omega}_{0,(1,1)}$	1.147421731	1.147598493	0.980984
$0.2\bar{\omega}_{0,(1,1)}$	1.147333371	1.147686911	0.819464
Mode (1,2)			
0	0.661285931	0.661285931	2.049864
$0.1\bar{\omega}_{0,(1,2)}$	0.661241071	0.661330968	1.941637
$0.2\bar{\omega}_{0,(1,2)}$	0.661203037	0.661371972	1.843994
Mode (1,3)			
0	0.404794834	0.404794834	3.349733
$0.1\bar{\omega}_{0,(1,3)}$	0.404774958	0.404814611	3.306232
$0.2\bar{\omega}_{0,(1,3)}$	0.404755166	0.404834476	3.257703
Mode (1,4)			
0	0.286536716	0.286536716	4.734211
$0.1\bar{\omega}_{0,(1,4)}$	0.286525590	0.286547877	4.706314
$0.2\bar{\omega}_{0,(1,4)}$	0.286514464	0.286559039	4.678707

Table 6.2

Unstable regions for the transverse modes of a simply supported isotropic rotating cylindrical shell of $v = 0.3$ and geometric properties $L/R = 2$ and $R/h = 100$ and subjected to compressive loading of $\eta_o = 0.1\eta_{cri}$.

$\bar{\Omega}$	p_1	p_2	$\Theta(\times 10^{-3})$
Mode (1,1)			
0	1.142776005	1.142776005	1.186329
$0.1\bar{\omega}_{0,(1,1)}$	1.142687959	1.142864039	0.986486
$0.2\bar{\omega}_{0,(1,1)}$	1.142599903	1.142952053	0.825193
Mode (1,2)			
0	0.653034441	0.653034441	2.077394
$0.1\bar{\omega}_{0,(1,2)}$	0.652993496	0.653075391	1.977545
$0.2\bar{\omega}_{0,(1,2)}$	0.652952541	0.653116329	1.882734
Mode (1,3)			
0	0.391165368	0.391165368	3.473003
$0.1\bar{\omega}_{0,(1,3)}$	0.391146213	0.391184531	3.425658
$0.2\bar{\omega}_{0,(1,3)}$	0.391127076	0.391203710	3.378731
Mode (1,4)			
0	0.266927704	0.266927704	5.102492
$0.1\bar{\omega}_{0,(1,4)}$	0.266917338	0.266938101	5.076625
$0.2\bar{\omega}_{0,(1,4)}$	0.266906997	0.266948523	5.050886

Table 6.3

Unstable regions for the transverse modes of a simply supported isotropic rotating cylindrical shell of $v = 0.3$ and geometric properties $L/R = 2$ and $R/h = 100$ and subjected to extensional (tensile) loading of $\eta_o = 0.2\eta_{cri}$.

$\bar{\Omega}$	p_1	p_2	$\Theta(\times 10^{-3})$
Mode (1,1)			
0	1.149869661	1.149869661	2.357024
$0.1\bar{\omega}_{0,(1,1)}$	1.149781109	1.149958277	2.147079
$0.2\bar{\omega}_{0,(1,1)}$	1.149692491	1.150046828	1.956532
Mode (1,2)			
0	0.665372888	0.665372888	4.074289
$0.1\bar{\omega}_{0,(1,2)}$	0.665331190	0.665414637	3.976419
$0.2\bar{\omega}_{0,(1,2)}$	0.665289466	0.665456352	3.876344
Mode (1,3)			
0	0.411439402	0.411439402	6.590848
$0.1\bar{\omega}_{0,(1,3)}$	0.411419248	0.411459555	6.541257
$0.2\bar{\omega}_{0,(1,3)}$	0.411399132	0.411479739	6.490882
Mode (1,4)			
0	0.295852732	0.295852732	9.169040
$0.1\bar{\omega}_{0,(1,4)}$	0.295841236	0.295864249	9.140877
$0.2\bar{\omega}_{0,(1,4)}$	0.295829786	0.295875810	9.112268

Table 6.4

Unstable regions for the transverse modes of a simply supported isotropic rotating cylindrical shell of $v = 0.3$ and geometric properties $L/R = 2$ and $R/h = 100$ and subjected to compressive loading of $\eta_o = 0.2\eta_{cri}$.

$\bar{\Omega}$	p_1	p_2	$\Theta(\times 10^{-3})$
Mode (1,1)			
0	1.140401294	1.140401294	2.378315
$0.1\bar{\omega}_{0,(1,1)}$	1.140313465	1.140489176	2.168811
$0.2\bar{\omega}_{0,(1,1)}$	1.140225612	1.140577031	1.979382
Mode (1,2)			
0	0.648868879	0.648868879	4.184567
$0.1\bar{\omega}_{0,(1,2)}$	0.648828206	0.648909577	4.084123
$0.2\bar{\omega}_{0,(1,2)}$	0.648787533	0.648950279	3.986121
Mode (1,3)			
0	0.384168422	0.384168422	7.074257
$0.1\bar{\omega}_{0,(1,3)}$	0.384149608	0.384187240	7.014284
$0.2\bar{\omega}_{0,(1,3)}$	0.384130813	0.384206074	6.967822
Mode (1,4)			
0	0.256559852	0.256559852	10.667037
$0.1\bar{\omega}_{0,(1,4)}$	0.256549877	0.256569833	10.554579
$0.2\bar{\omega}_{0,(1,4)}$	0.256539937	0.256579849	10.442165

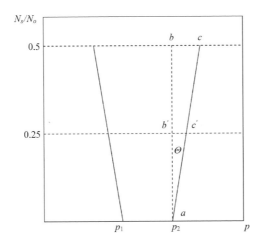

Figure 6.7 An unstable region in the N_s/N_o–p plane.

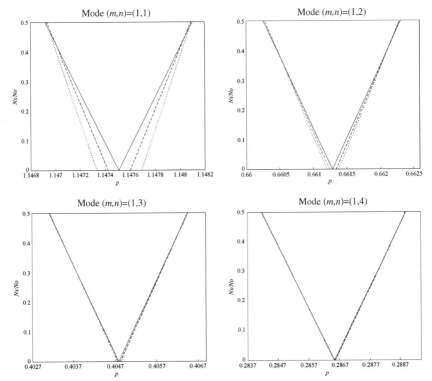

Figure 6.8 Unstable regions for the transverse modes of a simply supported isotropic rotating cylindrical shell of $v = 0.3$ and geometric properties $L/R = 2$ and $R/h = 100$ and subjected to extensional (tensile) loading of $\eta_o = 0.1\eta_{cri}$. — $\bar{\Omega} = 0$; --- $\bar{\Omega} = 0.1\bar{\omega}_o$; ····· $\bar{\Omega} = 0.2\bar{\omega}_o$.

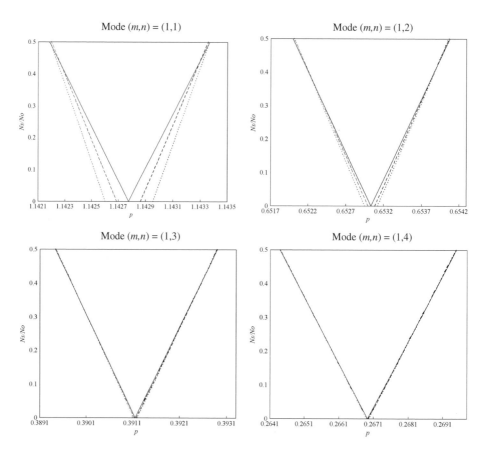

Figure 6.9 Unstable regions for the transverse modes of a simply supported isotropic rotating cylindrical shell of $v = 0.3$ and geometric properties $L/R = 2$ and $R/h = 100$ and subjected to compressive loading of $\eta_0 = -0.1\eta_{\text{cri}}$. — $\bar{\Omega} = 0$; --- $\bar{\Omega} = 0.1\bar{\omega}_o$; ...... $\bar{\Omega} = 0.2\bar{\omega}_o$.

p is defined as

$$p = RP\left(\frac{\rho(1-v^2)}{E}\right)^{1/2} \tag{6.35}$$

Each unstable region is bounded by two lines which may or may not originate from a common point from the p axis. The two curves appear at first glance to be straight lines but are, in fact, two very slight "outward" curving plots. For the sake of tabular presentation, each unstable region is defined by its two originating points, p_1 and p_2, from the p axis with $\eta_s = 0$. If the two curves originate from the same point, as is the case

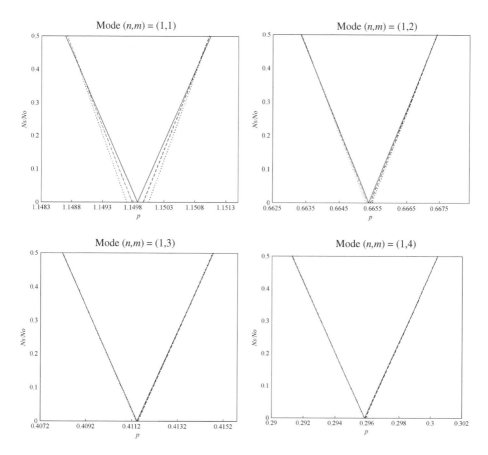

Figure 6.10 Unstable regions for the transverse modes of a simply supported isotropic rotating cylindrical shell of $v = 0.3$ and geometric properties $L/R = 2$ and $R/h = 100$ and subjected to extensional (tensile) loading of $\eta_o = 0.2\eta_{\mathrm{cri}}$. — $\bar{\Omega} = 0$; --- $\bar{\Omega} = 0.1\bar{\omega}_o$; ······ $\bar{\Omega} = 0.2\bar{\omega}_o$.

for the non-rotating shell, then $p_1 = p_2$. The angle subtended, Θ, is also introduced. It is calculated based on the arctangent of the right-angled triangle, abc, as shown in Fig. 6.7. This angle gives an accurate measure of the slope of the boundary of the unstable region as calculations done with the smaller similar triangle, $ab'c'$ (see Fig. 6.7), are within 0.1%.

The results presented in this study are for a simply supported isotropic rotating cylindrical shell with Poisson's ratio $v = 0.3$, and geometric properties $L/R = 2$ and $R/h = 100$. The modes of interest here are the transverse modes and the two higher axial and circumferential modes are neglected in the analysis. Results presented are for different rotational speeds for the transverse modes of modes (1,1), (1,2), (1,3) and

(1,4), respectively. As in the preceding sub-section on critical speed, the dynamic stability results presented here exclude those for circumferential wave number $n > 4$ due to the limitation of Donnell's equations to the higher modes for short to moderate length shells.

The values of η_o are chosen to be in terms of η_{cri}, which is the critical buckling load of a simply supported circular cylindrical shell subjected to static compressive axial load and is given by

$$\eta_{cri} = N_{cri} \frac{1 - v^2}{Eh} \tag{6.36}$$

where N_{cri} as given by Timoshenko & Gere [1961] is

$$N_{cri} = \frac{Eh^2}{R\sqrt{3(1 - v^2)}} \tag{6.37}$$

and since the Poisson's ratio is taken as $v = 0.3$, therefore,

$$\eta_{cri} = 0.5507 \frac{h}{R} \tag{6.38}$$

Table 6.1 gives the tabular representations for Fig. 6.8 which are results for the case of tensile loading $\eta_o = 0.1\eta_{cri}$. Corresponding results for compressive loading of $\eta_o = -0.1\eta_{cri}$ are given in Table 6.2 and Fig. 6.9. The corresponding results for increased loading magnitudes are given in Table 6.3 and Fig. 6.10 for tensile loading of $\eta_o = 0.2\eta_{cri}$, and in Table 6.4 and Fig. 6.11 for compressive loading of $\eta_o = -0.2\eta_{cri}$.

The nondimensional rotational speeds, $\bar{\Omega}$, used for each mode are in terms of the dimensionless natural frequencies of the non-rotating shell, $\bar{\omega}_o$, of that particular mode and under corresponding tensile loading. The natural frequency of the shell, ω, and the rotational speed, Ω, are nondimensionalized in the same way as the excitation frequency p in Eq. (6.35).

It is observed from the results presented that the introduction of rotation causes the two boundaries of the unstable regions to mutually shift apart. Therefore, unlike the case of a stationary shell where the unstable regions originate from a single point on the abscissa, the unstable regions now originate from two points (p_1 and p_2) on the abscissa in the presence of rotation. Furthermore, it is also observed that as the rotational speeds increase, the boundaries of each unstable region shift further and further away from each other and the region broadens. In some of the figures, this phenomenon is not immediately apparent, as the boundaries of the unstable regions for these cases have just

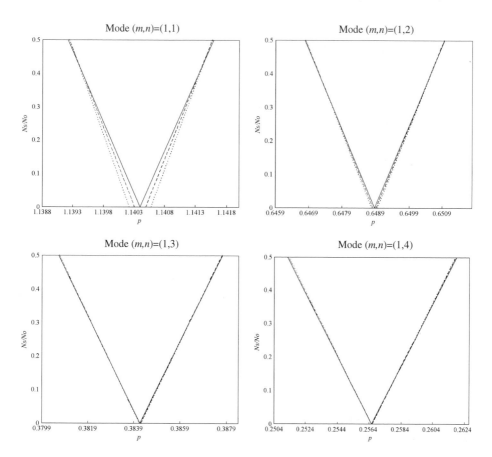

Figure 6.11 Unstable regions for the transverse modes of a simply supported isotropic rotating cylindrical shell of $v = 0.3$ and geometric properties $L/R = 2$ and $R/h = 100$ and subjected to compressive loading of $\eta_o = -0.2\eta_{\text{cri}}$. — $\bar{\Omega} = 0$; --- $\bar{\Omega} = 0.1\bar{\omega}_o$; ⋯⋯ $\bar{\Omega} = 0.2\bar{\omega}_o$.

begun to shift away from each other. However, the tabular results clearly show the presence of this phenomenon. As the Coriolis terms are proportional to the rotational speed, it can be concluded that the Coriolis effects destabilize the rotating shell causing the widths of the unstable regions to increase. Based on the relative magnitudes of the above-mentioned shifts in the unstable region boundaries, it can also be concluded that for the present rotating shell configuration, the lower modes of (1,1) and (1,2) are more sensitive to the Coriolis effects than the higher modes of (1,3) and (1,4).

From the results, it is observed that as the magnitude of the tensile loading is increased, the unstable regions shift to the right having higher points of origins. The converse is true when the magnitude of the compressive loading is increased. This can be expected as the natural frequencies of a shell increase as it is axially stretched and decrease as it is compressed. The sizes of the unstable regions in this study are now dependent upon two variables, firstly the $p_2 - p_1$ difference and secondly the subtended angle Θ. From the plotted figures, it is observed that as the magnitude of the axial loading is increased for both tensile and compressive cases, the sizes of the unstable regions also generally increase. It is also of interest to note that when comparing tensile and compressive loadings of the same magnitude, the $p_2 - p_1$ differences of the unstable regions associated with the tensile loadings are generally larger, while the subtended angles Θ of the unstable regions associated with the compressive loadings are generally larger.

REFERENCES

A.E. Armenakas and G. Herrmann. (1963). Vibrations of infinitely long cylindrical shells under initial stress. *AIAA Journal*, 1, 100-106.

M. Bacon and C.W. Bert. (1967). Unsymmetric free vibrations of orthotropic sandwich shells of revolution. *AIAA Journal*, 5, 413-417.

R.E. Bellman and J. Casti. (1971). Differential quadrature and long-term integration. *Journal of Mathematics and Analytical Applications*, 34, 235-238.

R.E. Bellman, B.G. Kashef and J. Casti. (1972). Differential quadrature: a technique for the rapid solution of nonlinear partial differential equations. *Journal of Computational Physics*, 10, 40-52.

C.W. Bert and M. Mailk. (1996a). Free vibration analysis of tapered rectangular plates by differential quadrature method: a semi-analytical approach. *Journal of Sound and Vibration*, 190, 41-63.

C.W. Bert and M. Mailk. (1996b). On the relative effects of transverse shear deformation and rotary inertia on the free vibration of symmetric cross-ply laminated plates. *Journal of Sound and Vibration*, 193, 927-933.

C.W. Bert and M. Mailk. (1996c). Semianalytical differential quadrature solution for free vibration analysis of rectangular plates. *AIAA Journal*, 34, 601-606.

C.W. Bert and M. Mailk. (1996d). Free vibration analysis of thin cylindrical shells by the differential quadrature method. *Journal of Pressure Vessel Technology*, 118, 1-12.

C.W. Bert and M. Mailk. (1996e). The differential quadrature method for irregular domains and application to plate vibration. *International Journal of Mechanical Sciences*, 38, 589-606.

C.W. Bert and M. Mailk. (1996f). Differential quadrature method in computational mechanics: a review. *ASME Applied Mechanics Reviews*, 49, 1-28.

C.W. Bert, X. Wang and A.G. Striz. (1993). Differential quadrature for static and free vibration analyses of anisotropic plates. *International Journal of Solids and Structures*, 30, 1737-1744.

C.W. Bert, X. Wang and A.G. Striz. (1994). Convergence of the DQ method in the analysis of anisotropic plates. *Journal of Sound and Vibration*, 170, 140-144.

H.H. Bleich and M.L. Baron. (1954). Free and forced vibrations of an infinitely long cylindrical shell in an infinite acoustic medium. *Journal of Applied Mechanics*. June.

V.V. Bolotin. (1964). *The Dynamic Stability of Elastic Systems*. San Francisco: Holden-Day.

G.H. Bryan. (1890). On the beats in the vibration of revolving cylinder or bell. *Proceedings of the Cambridge Philosophical Society*, 7, 101-111.

Z. Brzoska. (1953). Critical speeds of short drums. *Tech. Lotn*, 8, 151-156. (In Polish).

X.X. Cai. (1994). Free vibration of a thin rotating shell of revolution. *Computers and Structures*, 53, 155-160.

Y.B. Chang, T.Y. Yang and W. Soedel. (1983). Linear dynamic analysis of revolutional shells using finite elements and modal expansion. *Journal of Sound and Vibration*, 86, 523-538.

Y. Chen, H.B. Zhao, Z.P. Shen, I. Grieger and B.H. Kröplin. (1993). Vibrations of high speed rotating shells with calculations for cylindrical shells. *Journal of Sound and Vibration*, 160, 137-160.

D.K. Chun and C.W. Bert. (1993). Critical speed analysis of laminated composite, hollow drive shafts. *Composites Engineering*, 3, 633-643.

H. Chung. (1981). Free vibration analysis of circular cylindrical shells. *Journal of Sound and Vibration*, 74, 331-350.

R.A. DiTaranto and M. Lessen. (1964). Coriolis acceleration effect on the vibration of a rotating thin-walled circular cylinder. *ASME Journal of Applied Mechanics*, 31, 700-701.

S.B. Dong. (1977). A block-stodola eigensolution technique for large algebraic systems with non-symmetrical matrices. *International Journal for Numerical Methods in Engineering*, 11, 247-267.

L.H. Donnell. (1933). Stability of thin walled tubes in torsion. *NACA Report No. 479*.

H.L.M. dos Reis, R.B. Goldman and P.H. Verstrate. (1987). Thin-walled laminated composite cylindrical tubes: Part III — critical speed analysis. *Journal of Composites Technology and Research*, 9, 58-62.

H. Du, M.K. Lim and R.M. Lin. (1994). Application of generalized differential quadrature method to structural problems. *International Journal for Numerical Methods in Engineering*, 37, 1881-1896.

H. Du, M.K. Lim and R.M. Lin. (1995). Application of generalized differential quadrature method to vibration analysis. *Journal of Sound and Vibration*, 181, 279-293.

C.L. Dym. (1973). Some new results for the vibrations of circular cylinders. *Journal of Sound and Vibration*, 29, 189-205.

W. Flügge. (1960). *Stresses in Shells*. Berlin: Springer.

K. Forsberg. (1964). Influence of boundary conditions on the modal characteristics of thin cylindrical shells. *AIAA Journal*, 2, 2150-2157.

C.H.J. Fox and D.J.W. Hardie. (1985). Harmonic response of rotating cylindrical shells. *Journal of Sound and Vibration*, 101, 495-510.

Y.C. Fung, E.E. Sechler and A. Kaplan. (1957). On the vibration of thin cylindrical shells under internal pressure. *Journal of the Aeronautical Sciences*. September.

G. Galileo. (1939). *Dialogue Concerning Two New Sciences (1638)*. Evanston: Northwestern University Press.

R. Grybos. (1961). The problem of stability of a rotating elastic solid. *Archiwum Mechaniki Stosowanej*, 13, 761-774.

D. Guo, F.L. Chu and Z.C. Zheng. (2001). The influence of rotation on vibration of a thick cylindrical shell. *Journal of Sound and Vibration*, 242, 487-505.

R.H. Gutierrez and P.A.A. Laura. (1995). Analysis of vibrating, thin, rectangular plates with point supports by the method of differential quadrature. *Ocean Engineering*, 22, 101-103.

R.H. Gutierrez, P.A.A. Laura and R.E. Ross. (1994). Method of differential quadrature and its application to the approximate solution of ocean engineering problems. *Ocean Engineering*, 21, 57-66.

D.M. Haughton. (1982). The vibration of rotating elastic membrane cylinders. *International Journal of Engineering Science*, 20, 835-844.

S.C. Huang. (1987). Effects of Coriolis acceleration on the free and forced vibrations of spinning structures. PhD Thesis, Purdue University.

S.C. Huang and B.S. Hsu. (1990). Resonant phenomena of a rotating cylindrical shell subjected to a harmonic moving load. *Journal of Sound and Vibration*, 136, 215-228.

S.C. Huang and B.S. Hsu. (1992). Vibration of spinning ring-stiffened thin cylindrical shells. *AIAA Journal*, 30, 2291-2298.

S.C. Huang and W. Soedel. (1987a). Effects of Coriolis acceleration on the free and forced in-plane vibrations of rotating rings on elastic foundation. *Journal of Sound and Vibration*, 115, 253-274.

S.C. Huang and W. Soedel. (1987b). Response of rotating rings to harmonic and periodic loading and comparison with the inverted problem. *Journal of Sound and Vibration*, 118, 253-270.

S.C. Huang and W. Soedel. (1988a). On the forced vibration of simply supported rotating cylindrical shells. *Journal of the Acoustical Society of America*, 84, 275-285.

S.C. Huang and W. Soedel. (1988b). Effects of Coriolis acceleration on the forced vibration of rotating cylindrical shells. *Journal of Applied Mechanics*, 55, 231-233.

T. Irie, G. Yamada and K. Tanaka. (1984). Natural frequencies of truncated conical shells. *Journal of Sound and Vibration*, 92, 447-453.

S.K. Jan, C.W. Bert and A.G. Striz. (1989). Application of differential quadrature to static analysis of structural components. *International Journal for Numerical Methods in Engineering*, 28, 561-577.

A. Kayran and J.R. Vinson. (1990). Free vibration analysis of laminated composite truncated circular conical shells. *AIAA Journal*, 28, 1259-1269.

A.A. Khdeir and J.N. Reddy. (1990). Influence of edge conditions on the modal characteristics of cross-ply laminated shells. *Computers and Structures*, 34, 817-826.

A.A. Khdeir, J.N. Reddy and D. Frederick. (1989). A study of bending, vibration and buckling of cross-ply circular cylindrical shells with various shell theories. *International Journal of Engineering Science*, 27, 1337-1351.

C.D. Kim and C.W. Bert. (1993). Critical speed analysis of laminated composite hollow drive shafts. *Composites Engineering*, 3, 633-643.

G. Kirchhoff. (1876) *Vorlesungen uber Mathematische Physik Mechanik*, 1.

Y. Kobayashi and G. Yamada. (1991). Free vibration of a spinning polar orthotropic shallow spherical shell. *JSME International Journal*, 34, 233-238.

B.L. Koff and Y.M. El-Aini. (1993). A simple frequency expression for the in-plane vibrations of rotating rings. *ASME Journal of Engineering for Gas Turbines and Power*, 115, 234-238.

T. Koga. (1988). Effects of boundary conditions on the free vibrations of circular cylindrical shells. *AIAA Journal*, 26, 1387-1394.

L.R. Koval and E.T. Cranch. (1962). On the free vibrations of thin cylindrical shells subjected to an initial static torque. *4th US National Congress on Applied Mechanics*, June 18–21, 1, pp. 107-117.

A.R. Kukreti, J. Farsa and C.W. Bert. (1996). Differential quadrature and Rayleigh–Ritz methods to determine the fundamental frequencies of simply supported rectangular plates with linearly varying thickness. *Journal of Sound and Vibration*, 189, 103-122.

S.S.E. Lam. (1993). Application of the differential quadrature method to two-dimensional problems with arbitrary geometry. *Computers and Structures*, 47, 459-464.

K.Y. Lam and Hua Li. (1997). Vibration analysis of a rotating truncated circular conical shell. *International Journal of Solids and Structures*, 34, 2183-2197.

K.Y. Lam and Hua Li. (1999a). Influence of boundary conditions on the frequency characteristics of a rotating truncated circular conical shell. *Journal of Sound and Vibration*, 223, 171-195.

K.Y. Lam and Hua Li. (1999b). On free vibration of a rotating truncated circular orthotropic conical shell. *Composites: Part B*, 30, 135-144.

K.Y. Lam and Hua Li. (2000a). Influence of initial pressure on frequency characteristics of a rotating truncated circular conical shell. *International Journal of Mechanical Sciences*, 42, 213-236.

K.Y. Lam and Hua Li. (2000b). Generalized differential quadrature for frequency of rotating multilayered conical shell. *ASCE Journal of Engineering Mechanics*, 126, 1156-1162.

K.Y. Lam, Hua Li, T.Y. Ng and C.F. Chua. (2002). Generalized differential quadrature method for the free vibration of truncated conical panels. *Journal of Sound and Vibration*, 251, 329-348.

K.Y. Lam and C.T. Loy. (1994). On vibrations of thin rotating laminated composite cylindrical shells. *Composites Engineering*, 4, 1153-1167.

K.Y. Lam and C.T. Loy. (1995a). Free vibrations of a rotating multi-layered cylindrical shell. *International Journal of Solids and Structures*, 32, 647-663.

K.Y. Lam and C.T. Loy. (1995b). Analysis of rotating laminated cylindrical shells by different thin shell theories. *Journal of Sound and Vibration*, 186, 23-35.

K.Y. Lam and C.T. Loy. (1998). Influence of boundary conditions for a thin laminated rotating cylindrical shell. *Composite Structures*, 41, 215-228.

K.Y. Lam and Q. Wu. (1999). Vibrations of thick rotating laminated composite cylindrical shells. *Journal of Sound and Vibration*, 225, 483-501.

K.Y. Lam, Z. Zong and Q.X. Wang. (2003). Dynamic response of a laminated pipeline on the seabed subjected to underwater shock. *Composites: Part B*, 34, 59-66.

A.W. Leissa. (1993). *Vibration of Shells*. New York, USA: Acoustical Society of America.

Hua Li. (2000a). Influence of boundary conditions on the free vibrations of rotating truncated circular multi-layered conical shells. *Composites: Part B*, 31, 265-275.

Hua Li. (2000b). Frequency characteristics of a rotating truncated circular layered conical shell. *Composite Structures*, 50, 59-68.

Hua Li. (2000c). Frequency analysis of rotating truncated circular orthotropic conical shells with different boundary conditions. *Composites Science and Technology*, 60, 2945-2955.

Hua Li and K.Y. Lam. (1998). Frequency characteristics of a thin rotating cylindrical shell using the generalized differential quadrature method. *International Journal of Mechanical Sciences*, 40, 443-459.

Hua Li and K.Y. Lam. (2000). The generalized differential quadrature method for frequency analysis of a rotating conical shell with initial pressure. *International Journal for Numerical Methods in Engineering*, 48, 1703-1722.

Hua Li and K.Y. Lam. (2001). Orthotropic influence on frequency characteristics of a rotating composite laminated conical shell by the generalized differential quadrature method. *International Journal of Solids and Structures*, 38, 3995-4015.

K.M. Liew, J.B. Han and Z.M. Xiao. (1996). Differential quadrature method for thick symmetric cross-ply laminates with first-order shear flexibility. *International Journal of Solids and Structures*, 33, 2647-2658.

A.E. Love. (1888). The small free vibrations and deformations of a thin elastic shell. *Philosophical Transactions of the Royal Society of London, Series A*, 179, 491-546.

A.E.H. Love. (1927). *A Treatise on the Mathematical Theory of Elasticity*, 4th edn. Cambridge: Cambridge University Press, pp. 528-530.

C.T. Loy, K.Y. Lam, Hua Li and G.R. Liu. (1999). Vibration of antisymmetric angle-ply laminated cylindrical panels with different boundary conditions. *Quarterly Journal of Mechanics and Applied Mathematics*, 52, 55-71.

H.J. Macke. (1966). Travelling wave vibration of gas turbine engine shell. *ASME Journal of Engineering for Power*, 88, 179-187.

A. Maewal. (1981). Nonlinear harmonic oscillations of gyroscopic structural systems and the case of a rotating ring. *ASME Journal of Applied Mechanics*, 48, 627-633.

M. Malik and C.W. Bert. (1995). Differential quadrature analysis of free vibration of symmetric cross-ply laminates with shear deformation and rotatory inertia. *Shock and Vibration*, 2, 321-338.

M. Malik and C.W. Bert. (1996). Implementing multiple boundary conditions in the DQ solution of higher-order PDE's: application to free vibration of plates. *International Journal for Numerical Methods in Engineering*, 39, 1237-1258.

S. Markus. (1988). The mechanics of vibrations of cylindrical shells. In *Studies in Applied Mechanics*, 17. Amsterdam: Elsevier.

R.D. Mindlin. (1951). Influence of rotatory inertia and shear on flexural motions of isotropic, elastic plates. *ASME Journal of Applied Mechanics*, 18, 31-38.

R.D. Mindlin and H. Deresiewicz. (1954). Thickness-shear and flexural vibrations of a circular disk. *Journal of Applied Physics*, 25, 1329-1332.

R.D. Mindlin, A. Schacknow and H. Deresiewicz. (1956). Flexural vibration of rectangular plates. *ASME Journal of Applied Mechanics*, 23, 430-436.

R. Miserentino and L.F. Vosteen. (1965). Vibration tests of pressurized thin-walled cylindrical shells. *NASA TN D* 3066.

K. Mizoguchi. (1963). Vibration of a rotating cylindrical shell. *Transactions of the Japan Society of Mechanical Engineers*, 29-203, 1217-1225 (In Japanese).

T.Y. Ng. (2003). Erratum to "Parametric resonance of a rotating cylindrical shell subjected to periodic axial loads" [*Journal of Sound and Vibration* (1998), 214, 513-529]. *Journal of Sound and Vibration*, 263, 705-708.

T.Y. Ng and K.Y. Lam. (1999). Vibration and critical speed of a rotating cylindrical shell subjected to axial loading. *Applied Acoustics*, 56, 273-282.

T.Y. Ng, K.Y. Lam and J.N. Reddy. (1998). Parametric resonance of a rotating cylindrical shell subjected to periodic axial loads. *Journal of Sound and Vibration*, 214, 513-529.

T.Y. Ng, Hua Li, K.Y. Lam and C.F. Chua. (2003). Frequency analysis of rotating conical panels: a generalized differential quadrature approach. *ASME Journal of Applied Mechanics*, 70, 601-605.

T.Y. Ng, Hua Li and K.Y. Lam. (2003). Generalized differential quadrature for free vibration of rotating composite laminated conical shell with various boundary conditions. *International Journal of Mechanical Sciences*, 45, 567-587.

T.Y. Ng, Hua Li, K.Y. Lam and C.T. Loy. (1999). Parametric instability of conical shells by the generalized differential quadrature method. *International Journal for Numerical Methods in Engineering*, 44, 819-837.

A. Nosier and J.N. Reddy. (1992). Vibration and stability analysis of cross-ply laminated circular cylindrical shells. *Journal of Sound and Vibration*, 157, 139-159.

V.V. Novozhilov. (1964). *Thin Shell Theory*, 2nd edn. The Netherlands: Noordhoff.

J. Padovan. (1973). Natural frequencies of rotating prestressed cylinders. *Journal of Sound and Vibration*, 31, 469-482.

J. Padovan. (1975a). Numerical analysis of asymmetric frequency and buckling eigenvalues of prestressed rotating anisotropic shells of revolution. *Computers and Structures*, 5, 145-154.

J. Padovan. (1975b). Traveling waves vibrations and buckling of rotating anisotropic shells of revolution by finite elements. *International Journal of Solids and Structures*, 11, 1367-1380.

L.E. Penzes and H. Kraus. (1972). Free vibration of prestresses cylindrical shells having arbitrary homogeneous boundary conditions. *AIAA Journal*, 10, 1309-1313.

J.R. Quan and C.T. Chang. (1989). New insights in solving distributed system equations by the quadrature method — I. analysis. *Computers and Chemical Engineering*, 13, 779-788.

O. Rand and Y. Stavsky. (1991). Free vibrations of spinning composite cylindrical shells. *International Journal of Solids and Structures*, 28, 831-843.

J.N. Reddy. (1984a). A simple higher-order theory for laminated composite plates. *ASME Journal of Applied Mechanics*, 51, 745-752.

J.N. Reddy. (1984b). *Energy and Variational Methods in Applied Mechanics — with an Introduction to the Finite Element Method*. New York: Wiley.

J.N. Reddy. (1986). *Applied Functional Analysis and Variational Methods in Engineering*. New York: McGraw-Hill.

J.N. Reddy. (1997). *Mechanics of Laminated Composite Plates — Theory and Analysis*. Boca Raton, FL: CRC Press.

J.N. Reddy. (1999). *Theory and Analysis of Elastic Plates*. Philadelphia, PA: Taylor and Francis.

J.N. Reddy. (2004). *Mechanics of Laminated Composite Plates and Shells — Theory and Analysis*. Boca Raton, FL: CRC Press.

E. Reissner. (1944). On the theory of bending of elastic plates. *Journal of Mathematical Physics*, 23, 184-191.

E. Reissner. (1952). Stress strain relations in the theory of thin elastic shells. *Journal of Mathematical Physics*, 31, 109-119.

C.T.F. Ross. (1975). Finite elements for the vibration of cones and cylinders. *International Journal for Numerical Methods in Engineering*, 9, 833-845.

T. Saito and M. Endo. (1986a). Vibration of finite length, rotating cylindrical shells. *Journal of Sound and Vibration*, 107, 17-28.

T. Saito and M. Endo. (1986b). Vibration analysis of rotating cylindrical shells based on the Timoshenko beam theory. *Bulletin of JSME*, 29, 1239-1245.

T. Saito and M. Endo. (1986c). The vibration of rotating cylindrical shells. *Bulletin of JSME*, 29, 3505-3509.

T. Saito, M. Endo and K. Fujimoto. (1989). Vibration analysis of thick rotating cylindrical shells based on the two-dimensional elasticity theory. *JSME International Journal*, 32, 585-591.

T. Saito, Y. Tsukahara and M. Endo. (1986). Vibration of rotating prestressed cylindrical shells. *Bulletin of JSME*, 29, 1572-1578.

J.L. Sanders. (1959). An improved first approximation theory for thin shells. *NASA Report NASA-TR-24*.

N. Sankaranarayanan, K. Chandrasekaran and G. Ramaiyan. (1987). Axisymmetric vibrations of laminated conical shells of variable thickness. *Journal of Sound and Vibration*, 118, 151-161.

C.B. Sharma. (1973). Frequencies of clamped-free circular cylindrical shell. *Journal of Sound and Vibration*, 30, 525-528.

G.J. Shevchuk and P. Thullen. (1978). Flexural vibrations of rotating electromagnetic shields. *ASME Journal of Mechanical Design*, 101, 133-137.

C. Shu (1991). Generalized differential–integral quadrature and application to the simulation of incompressible viscous flows including parallel computation. PhD Thesis, University of Glasgow.

C. Shu, B.C. Khoo and K.S. Yeo. (1994). Numerical solutions of incompressible Navier–Stokes equations by generalized differential quadrature. *Finite Elements in Analysis and Design*, 18, 83-97.

C. Shu and B.E. Richard. (1992). Application of generalized differential quadrature to solve two-dimensional incompressible Navier–Stokes equations. *International Journal for Numerical Methods in Fluids*, 15, 791-798.

K.R.Y. Simha, R. Jain and K. Ramachandra. (1994). Variable density approach for rotating shallow shell of variable thickness. *International Journal of Solids and Structures*, 31, 849-863.

A.V. Singh, L. Zhu and S. Mirza. (1991). On asymmetric vibrations of layered orthotropic shells of revolution. *Journal of Sound and Vibration*, 148, 265-277.

K.R. Sivadas. (1995). Vibration analysis of pre-stressed rotating thick circular conical shell. *Journal of Sound and Vibration*, 186, 99-109.

K.R. Sivadas and N. Ganesan. (1991). Vibration analysis of laminated conical shells with variable thickness. *Journal of Sound and Vibration*, 148, 477-491.

K.R. Sivadas and N. Ganesan. (1992). Vibration analysis of thick composite clamped conical shells of varying thickness. *Journal of Sound and Vibration*, 152, 27-37.

K.R. Sivadas and N. Ganesan. (1993). Axisymmetric vibration analysis of thick cylindrical shell with variable thickness. *Journal of Sound and Vibration*, 160, 387-400.

K.R. Sivadas and N. Ganesan. (1994). Effect of rotation on vibration of moderately thick circular cylindrical shells. *ASME Journal of Vibration and Acoustics*, 116, 198-202.

A. Smirnov. (1989). Free vibrations of the rotating shells of revolution. *ASME Journal of Applied Mechanics*, 56, 423-429.

W. Soedel. (1993). *Vibrations of Shells and Plates*, 2nd edn. New York: Marcel Dekker.

K.P. Soldatos. (1984). A comparison of some shell theories used for the dynamic analysis of cross-ply laminated circular cylindrical panels. *Journal of Sound and Vibration*, 97, 305-319.

A.V. Srinivasan and G.F. Lauterbach. (1971). Travelling waves in rotating cylindrical shells. *ASME Journal of Engineering for Industries*, 93, 1229-1232.

R.S. Srinivasan and P.A. Krishnan. (1989). Dynamic analysis of stiffened conical shell panels. *Computers and Structures*, 33, 831-837.

A.G. Striz, X. Wang and C.W. Bert. (1995). Harmonic differential quadrature method and applications to analysis of structural components. *Acta Mechanica*, 111, 85-94.

B.S.K. Sundarasivarao and N. Ganesan. (1991). Deformation of varying thickness of conical shells subjected to axisymmetric loading with various end conditions. *Engineering Fracture Mechanics*, 39, 1003-1010.

K. Suzuki, T. Kosawada, G. Shikanai and K. Hayashi. (1993). Vibrations of rotating circular cylindrical shells with varying thickness. *Journal of Sound and Vibration*, 166, 267-282.

K. Suzuki, R. Takahashi and T. Kosawada. (1991). Analysis of vibrations of rotating thin circular cylindrical shells. *JSME International Journal*, 34, 19-25.

D.P. Thambiratnam and Y. Zhuge. (1993). Axisymmetric free vibration analysis of conical shells. *Engineering Structures*, 15, 83-89.

T. Timarci and K.P. Soldatos. (1995). Comparative dynamic studies for symmetric cross-ply circular cylindrical shells on the basis of a unified shear deformable shell theory. *Journal of Sound and Vibration*, 187, 609-624.

L.Y. Tong. (1993a). Free vibration of orthotropic conical shells. *International Journal of Engineering Science*, 31, 719-733.

L.Y. Tong. (1993b). Free vibration of composite laminated conical shells. *International Journal of Mechanical Sciences*, 35, 47-61.

X. Wang and C.B. Bert. (1993). A new approach in applying differential quadrature to static and free vibrational analyses of beams and plates. *Journal of Sound and Vibration*, 162, 566-572.

S.S. Wang and Y. Chen. (1974). Effects of rotation on the vibrations of circular cylindrical shells. *Journal of the Acoustical Society of America*, 55, 1340-1342.

J.T.S. Wang and C.C. Lin. (1993). Stresses in rotating composite cylindrical shells. *Composite Structures*, 25, 157-164.

H. Wang, K. Williams and W. Guan. (1998). A vibrational mode analysis of free finite-length thick cylinders using the finite element method. *ASME Journal of Vibration and Acoustics*, 120, 371-377.

V.I. Weingarten. (1966). The effect of internal or external pressure on the free vibrations of conical shells. *International Journal of Mechanical Sciences*, 8, 115-124.

H. Zinberg and M.F. Symonds. (1970). The development of advanced composite tail rotor driveshaft, *26th Annual Forum of the American Helicopter Society*, pp. 1-14.

A. Zohar and J. Aboudi. (1973). The free vibrations of thin circular finite rotating cylinder. *International Journal of Mechanical Sciences*, 15, 269-278.

SUBJECT INDEX

acceleration:
    centrifugal 2, 53, 106, 145
    coriolis 2, 4, 28, 53, 106, 110, 145, 203, 218,
        235
    relative 106
assumed-mode method 4, 5, 92, 113,
    137, 138, 141, 142

beam function 4, 5, 33–36, 39, 42, 45
bifurcation:
    natural frequencies 2, 45, 53, 234
    instability regions 5, 229, 230, 232, 238,
        239, 242
Bolotin's method 5, 230, 238
boundary condition:
    clamped 34, 39, 40, 42, 75
    free 30, 34, 39, 75, 97, 214, 218, 219
    geometric 35, 36
    simply supported 34, 39, 41, 42, 45,
        47, 48, 55, 67, 72, 75, 86, 87,
        138, 144, 145, 151, 171, 172,
        191, 205, 210
boundary value problem 24, 25, 75

centripetal force 19, 22
critical speed 3, 5, 25, 229, 230, 232, 234, 235,
    239, 242, 248
composite shell:
    cross-ply 135, 161, 204
    laminated 8, 15, 16, 75, 87, 160
    multi-layered 11, 12, 14, 47, 48,
        53, 57, 67
    orthotropic 43, 134, 150, 177
    sandwich-type 16, 28, 146, 150, 169–173,
        192, 194
convergence characteristics 39, 137, 141, 142

coordinate:
    curvilinear 16, 17, 24, 29, 92, 94
    global geometric 12
    material principal 12
    orthogonal 29, 92, 202
    principal 12, 18
curvature 8, 11, 17, 201, 215, 216

direction:
    axial 35, 217
    circumferential 12, 86, 115, 191, 213, 215
    longitudinal 35
    meridional 91, 115, 139, 140
displacement field 34, 36, 37, 114, 115, 141,
    203, 205, 206, 230
dynamic stability 5, 25, 229–232, 235, 239,
    242, 248

eigen:
    solution 38, 39, 132, 141, 207
    value 4, 24, 37–39, 130–133, 206, 242
    vector 24

frequency:
    circular 35, 37, 38, 41, 43, 56, 57, 87, 206,
        207, 220–222, 227, 248
    fundamental 54, 57, 67, 75, 210
    natural 2, 3, 28, 39, 53, 67, 76, 91, 115, 130,
        201, 202, 209, 214, 222, 229, 234, 235,
        239, 248, 250
first-order shear deformation theory (FSDT) 7,
    10, 201

Galerkin method 4, 33, 37, 39–41, 45, 92, 143
generalized differential quadrature (GDQ) 4,
    92, 113

governing equations for a rotating shell 12, 16, 24, 28, 32, 34, 37, 45, 104, 108, 128

Hooke's law 11, 12

inertia:
  force 19
  rotary 3, 5, 28, 201, 218
infinitesimal body 4, 18, 19
initial:
  hoop tension 30, 32, 110, 230, 231
  stress 5, 27, 92, 106, 107, 110, 179, 180
  pressure 107, 149, 155, 156, 158–160, 180–189, 194
isotropic 1, 3, 15, 27, 35, 39, 42, 43, 45, 53, 57, 67, 69, 86, 91, 106, 108, 134, 138–140, 142, 143, 148, 150, 155, 159, 180, 188, 193, 207, 219, 229, 247

Kirchhoff-Love hypotheses 7, 8, 47, 201

Lamé parameter 16, 18, 29, 92, 94

relations:
  constitutive 14, 34, 47, 52, 113, 161, 204
  geometric 8, 31, 34, 113
  kinematics 8, 218
  strain-displacement 8–12, 18, 31, 47, 103, 201, 203, 218
  stress-strain 4, 11, 12
resultants:
  force 13, 21
  moment 13, 21, 32, 201, 203–205

shear correction factor 201, 207, 210
shell:
  cylindrical 1, 3–5, 7–10, 14, 16, 24, 25, 27–79, 81, 82, 84–88, 93–97, 106, 138–141, 150, 180, 201, 202,
    204–214, 216–220, 229, 230, 232, 234, 239–249
  conical 1, 4, 10, 16, 24, 25, 39, 42, 43, 91, 92, 94–96, 98–101, 103, 104, 106–108, 111–115, 117, 118, 120–126, 128, 129, 132–143, 145, 146, 148, 150–183, 185, 187–194, 202
  spherical 1, 254
  stationary 2, 53, 54, 191, 219, 248
shell theory:
  Donnell 32, 33, 47–58, 229, 230
  Flügge 13, 16, 21, 27, 31, 39, 47–49, 52
  Love 8, 28, 34, 43, 47–49, 52
  Mindlin 5, 202, 207–209
  Novozhilov 18, 19
  Sanders 47, 52
  thick 28, 201, 211, 218
  thin 3, 4, 7, 8, 12, 13, 17, 27, 28, 47, 49, 50, 52, 201, 211, 229, 230
stacking sequence 43, 47, 55, 75
stiffness:
  bending 14, 15, 204
  coupling 14, 15, 204
  extensional 41, 56, 57, 76, 87
  tensile 14, 15
surface:
  middle 1, 17, 29, 75, 92, 202, 203
  reference 1, 7, 8, 11, 12, 16, 28, 75, 92, 171, 202
  undeformed 216, 217

tracer 4, 33
transverse shear:
  strain 91
  stress 7

vibration mode:
  axial 35, 239
  bending 215–217, 219, 222, 225, 227
  circumferential 35, 84, 215, 217, 221, 227, 239, 247

extensional 215, 225, 226
longitudinal 84, 214–217, 225, 226
radial 214, 217, 220, 221, 227
shear 214, 217, 220, 221, 227
torsional 216, 217, 226

wave:
   backward 39, 45, 49, 54–56, 67, 73, 75, 89,
      132, 133, 136, 141, 167, 170, 176, 178,
      187, 190, 196, 210, 211, 220, 222, 225,
      227
   forward 2, 39, 43, 45, 47, 53–58, 67, 86, 89,
      133, 136, 141, 145, 148, 149, 151–159,
      163–170, 175, 176, 178, 183–187,
      189, 190, 192–195, 210, 220–222,
      225, 227
   traveling 27, 39, 45, 49–51, 80, 81, 83, 86, 207
   standing 2, 39, 45, 47, 53, 86, 133, 145,
      162, 163, 173, 183, 191, 232
wave number:
   axial 35
   circumferential 35, 39, 40, 47, 49–54, 56,
      73, 75–79, 84, 86, 92, 115, 132,
      145–149, 151, 155, 159, 161–167,
      169, 172–175, 179, 181–187,
      189–191, 194, 195, 202, 206,
      208–211, 213, 248
   longitudinal 39, 58, 84, 86, 206, 213
   meridional 92, 132, 190, 191